T0212162

The Standardization of American Schooling

SECONDARY EDUCATION IN A CHANGING WORLD

*Series editors:* Barry M. Franklin and Gary McCulloch

Published by Palgrave Macmillan:

*The Comprehensive Public High School: Historical Perspectives*
By Geoffrey Sherington and Craig Campbell
(2006)

*Cyril Norwood and the Ideal of Secondary Education*
By Gary McCulloch
(2007)

*The Death of the Comprehensive High School?:*
*Historical, Contemporary, and Comparative Perspectives*
Edited by Barry M. Franklin and Gary McCulloch
(2007)

*The Emergence of Holocaust Education in American Schools*
By Thomas D. Fallace
(2008)

*The Standardization of American Schooling:*
*Linking Secondary and Higher Education, 1870–1910*
By Marc A. VanOverbeke
(2008)

# THE STANDARDIZATION OF AMERICAN SCHOOLING

## LINKING SECONDARY AND HIGHER EDUCATION, 1870–1910

MARC A. VANOVERBEKE

First published in 2008 by
PALGRAVE MACMILLAN™
175 Fifth Avenue, New York, N.Y. 10010 and
Houndmills, Basingstoke, Hampshire, England RG21 6XS
Companies and representatives throughout the world.

PALGRAVE MACMILLAN is the global academic imprint of the Palgrave Macmillan division of St. Martin's Press, LLC and of Palgrave Macmillan Ltd. Macmillan® is a registered trademark in the United States, United Kingdom and other countries. Palgrave is a registered trademark in the European Union and other countries.

ISBN 978-1-349-37356-7      ISBN 978-0-230-61259-4 (eBook)
DOI 10.1057/9780230612594

Library of Congress Cataloging-in-Publication Data

VanOverbeke, Marc A.
    The standardization of American schooling : linking secondary and higher education, 1870–1910 / by Marc A. VanOverbeke.
        p. cm.—(Secondary education in a changing world series)

    1. Education, Secondary—Aims and objectives—United States—History. 2. Education, Higher—United States—History.
3. Educational change—United States—History. I. Title.

LA222.V363 2008
373.9309—dc22                              2007041091

A catalogue record for this book is available from the British Library.

Design by Newgen Imaging Systems (P) Ltd., Chennai, India.

First edition: June 2008

*To My Family*

# Contents

# Series Editors' Preface

Among the educational issues affecting policymakers, public officials, and citizens in modern, democratic, and industrial societies, none has been more contentious than the role of secondary schooling. In establishing the Secondary Education in a Changing World series with Palgrave Macmillan, the intent is to provide a venue for scholars in different national settings to explore critical and controversial issues surrounding secondary education. The series will be a place for the airing and, hopefully, resolution of these controversial issues.

More than a century has elapsed since Emile Durkheim argued the importance of studying secondary education as a unity, rather than in relation to the wide range of subjects and the division of pedagogical labor of which it was composed. Only thus, he insisted, would it be possible to have the ends and aims of secondary education constantly in view. The failure to do so accounted for a great deal of the difficulty with which secondary education was faced. First, it meant that secondary education was "intellectually disorientated," between "a past which is dying and a future which is still undecided," and as a result "lacks the vigor and vitality which it once possessed." Second, the institutions of secondary education were not understood adequately in relation to their past, which was "the soil which nourished them and gave them their present meaning, and apart from which they cannot be examined without a great deal of impoverishment and distortion." And third, it was difficult for secondary school teachers who were responsible for putting policy reforms into practice to understand the nature of the problems and issues that prompted them.[1]

In the early years of the twenty-first century, Durkheim's strictures still have resonance. The intellectual disorientation of secondary education is more evident than ever as it is caught up in successive waves of policy changes. The connections between the present and the past have become increasingly hard to trace and untangle. Moreover, the distance between policymakers on the one hand and practitioners on the other has rarely seemed as immense as it is today. The key mission of the current series of books is, in the spirit of Durkheim, to address these underlying dilemmas of secondary education and to play a part in resolving them.

*The Standardization of American Schooling* is Marc VanOverbeke's account of the efforts of a host of late-nineteenth and early-twentieth-century educational reformers, including university presidents, professors, secondary school administrators, and teachers, to develop a system of accreditation to connect the work of the secondary school with that of the university. The key players in his story are University of Michigan president James B. Angell and Harvard University president Charles W. Eliot. Focusing largely but not exclusively on events at the University of Michigan, VanOverbeke examines the development of this system and its impact on both the high school and the university.

VanOverbeke begins his tale by placing both secondary and higher education in the context of late-nineteenth-century America and the social and economic changes then occurring. He notes that throughout much of this period there was not a clear dividing line between the work of colleges and secondary schools and both institutions competed with each other for students. This was not a situation that could continue as university administrators and faculty spoke increasingly of the unique role that higher education was coming to play in credentialing the nation's youth for middle-class life.

Beginning with Angell's efforts in Michigan to develop a system for inspecting and accrediting high schools, the volume explores the expansion of this system of accreditation to the Midwest, New England, and the South. In discussing the diffusion of the idea of accreditation across the nation, VanOverbeke points to the regional differences that affected what was becoming throughout the nation a hierarchical system of education. These developments had not only regional but national manifestations, and he includes a discussion of the relationships between these two spheres of activity.

The volume pays particular attention to the place of the high school in these developments. Here, VanOverbeke focuses his attention on the role of the National Education Association's Committee of Ten on Secondary School Studies under Eliot's leadership. The work of this committee, he argues, served to strike a balance between the interests of higher and secondary education. It defined the four-year high school course of study as the standard preparation for admission to college, while at the same time promoting a degree of flexibility into its course of study that made room in the university for the modern subjects that were coming to dominate the high school's curriculum. Overall, the book considers how and why during the late nineteenth and early twentieth centuries two separate educational institutions were able to develop a means of linking their work to create by the turn of the century an American system of education.

*The Standardization of American Schooling* emphasizes in a more definitive way than other previous volumes in this series the connection between secondary education and other educational institutions, in this case the university. As the series develops, we as series editors hope to highlight an array of such linkages and in so doing situate secondary education in the broader context in which it exists.

*Barry M. Franklin and Gary McCulloch*
*Series Editors*

# Note

1. Emile Durkheim, *The Evolution of Educational Thought: Lectures on the Formation and Development of Secondary Education in France*, trans. Peter Collins (London: Routledge and Kegan Paul, 1977), 8, 10.

# Acknowledgments

The process of shaping America's educational system occurred in different ways throughout the country. In building this system, educational reformers left wonderful, rich sources that I have used, but, as is the case with most historical research, the trail of these sources stretched from New England through the South and into the Midwest and West. In the process of tracking down these sources and writing this book, I benefited from the help, encouragement, and support of numerous friends, colleagues, librarians, and scholars. I have incurred many debts.

In particular, I want to acknowledge and thank the librarians and archivists I worked with at the University of Wisconsin–Madison, Northern Illinois University, the Harvard University Archives, the University of Chicago's Special Collections Research Center, and the University of Michigan's Bentley Historical Library—an exceptional place to do archival research. I would have been lost without the many individuals who found important documents buried deep within their archives and in other libraries throughout the country.

I was fortunate to receive funding at various stages that made it easier to travel to libraries, collect research, and write this book. In the early data-gathering stages of this project, Henry Lufler and the Wisconsin Center for the Advancement of Postsecondary Education (WISCAPE) at the University of Wisconsin–Madison offered me crucial support through a research assistantship. The Bentley Historical Library provided travel support through a Mark C. Stevens Researcher Travel Fellowship. A Spencer Foundation Dissertation Fellowship meant that I had the luxury of dedicating an entire year to writing the dissertation that became the basis for this book. As I continued to work on this project as an assistant professor, I received summer support from a Research and Artistry Award from Northern Illinois University and additional support from the university's College of Education and the Department of Leadership, Educational Psychology, and Foundations. I am grateful for all of this support, as I am for the encouragement of my colleagues at Northern Illinois University. They have made my first years as a professor richly rewarding ones.

Numerous individuals read portions of this manuscript and helped me work through problems and strengthen what I wanted to do. In particular, I want to thank James Baughman, Patricia Burch, Michael Fultz, Herbert Kliebard, and Adam Nelson. Barry Franklin and Gary McCulloch first showed an interest in publishing this project as part of their series on secondary education. I have enjoyed working with them and with Brigitte Shull, Amanda Johnson Moon, Kristy Lilas, and Julia Cohen at Palgrave Macmillan, who responded with patience to my many queries. I also want to acknowledge Philip Jordan, who in his undergraduate history courses at Hastings College first encouraged me to study history seriously as a scholar, and Rubén Donato at the University of Colorado–Boulder who showed me how exciting and enjoyable such work can be. William Reese was an exemplary mentor and advisor during my years as a graduate student at the University of Wisconsin–Madison, and Bill continues to be a trusted colleague and friend. I hope I can emulate his passion for the field and his superb teaching and scholarship.

Finally, this book is dedicated to my family, whose passion for education marked my childhood and continues to fascinate, amaze, and encourage me.

# Introduction

## I

In his first report to the Board of Regents as president of the University of Michigan in 1872, James B. Angell explained that his goal was "to raise the grade of our work as rapidly as the preparatory schools can raise theirs. We keep up constant communion with the superintendents and teachers to determine how far they can readily carry their students before they transfer them to us." Optimistically, he concluded, "They show a most praiseworthy desire to push up the scale of their work." Angell wanted to turn his institution into one of the nation's premier universities. To do that, he needed the support of the high schools and the well-educated students they might send up for advanced study.[1]

The president of the school board in Detroit agreed with the tenor of Angell's comments and told his board members that the high school course needed "some reconstruction, as the result of the recent changes in the standard of admission to the University."[2] This response must have heartened Angell, but not all secondary schools agreed so readily and transformed their standards in line with the university's curriculum. Nonetheless, both higher and secondary education were coming to understand in the late nineteenth century that their fortunes were inextricably linked.

What Angell wanted was an articulated system of education that aligned schools, courses, and standards at all levels so that students passed seamlessly from one grade to another and, importantly, from the secondary schools to the university. The development of this system of education shaped the colleges and universities and the secondary schools in profound ways. This book focuses on how Angell and other educators forged this articulated system of education, and considers why they created a standardized system at a time of rapid social and economic change. It traces the development of a peculiarly American educational structure, starting in 1870 with Michigan and Angell's attempts to articulate—or connect—higher and secondary education. Angell led the way in Michigan, but it was only an initial effort in a larger process that continues today. By the early twentieth century, when this study ends, however, most of the structure was in place. From Michigan, the story travels to other states, explores distinct regional differences, and considers some of the first, albeit less-than-successful, efforts to craft a standardized, efficient system of education on a national scale that matched the idealized efficiency of industrial America. Along the way, it introduces a host of pivotal figures, including university presidents, professors, secondary school administrators, and teachers, who were central in the campaign to restructure American education.

Unlike Germany and France—two countries that educators used as comparisons to highlight the weaknesses of American education—the United States did not have a centralized authority to direct and control this restructuring of American schools and colleges. The country's schools and universities were not part of an organic whole. Higher and secondary education for the most part grew independently of each other and possessed their own traditions and customs. Some secondary schools, especially in New England, aligned closely with that region's colleges and worked in harmony with them, but most colleges and secondary schools had little direct connection when Angell began his tenure as Michigan's president. Whatever relationship was to emerge had to come from the initiative and efforts of both levels working together. No state or federal agency possessed the authority to compel the creation of an efficient educational system that harmoniously linked higher and secondary education.

In the absence of any central authority to coordinate education, a group of professional educators and experts led the campaign to connect the secondary and higher schools so that students could move easily from one to the other. Usually consisting of university presidents and professors, as well as some reform-minded superintendents, principals, and school board members, this body of expert educators presented a striking contrast to local school boards and lay control. Their reforms were not always welcomed or easily implemented, and these reformers were never a completely unified group. They often disagreed on the appropriate course of action, and significant tensions between educators from the secondary schools and those from higher education repeatedly surfaced. However, they had to reconcile their conflicts and disagreements, and accommodate each other, since neither secondary nor higher education had sufficient power or authority to compel the other to embrace unwanted reforms.

These reformers may not have always agreed, but they profoundly altered the landscape of American education between 1870 and the early twentieth century. They created verifiable, recognizable standards for secondary schools, colleges, and universities throughout the nation, which helped to elevate the quality of work accomplished in these institutions, and they formalized the four-year high school as the standard route to college. What had been a flat or horizontal structure, where the secondary schools and colleges competed for students, was becoming hierarchical, with the high schools leading up to the colleges and the colleges funneling students to graduate programs and universities. This relationship highlighted the university as the pinnacle of American education, one with clear standards resting on a strong foundation of secondary schools and undergraduate colleges. While the secondary schools lost some of their standing in this new order, they gained greater authority for selecting who would go to college and for establishing the curriculum these students would study before entering college.

At the same time, however, these reforms undermined a tradition of local control that had guided elementary and secondary schools for decades. Reformers highlighted the importance of national standards and uniformity at a time when lay boards of education predominantly shaped the direction of secondary schools in line with local expectations, and many Americans regretted what was being lost. Although these reformers hoped to balance local needs with national norms and expectations, they ultimately promoted a system that tilted toward uniformity and standardization. While different leaders and programs marked articulation in various parts of the country, seemingly underscoring the importance of local traditions, these disparate initiatives eventually coalesced into a finely structured system that was, in many ways, national in scope.

A rich literature has furthered our understanding of education in this period, and this research is crucial to my study. But it has not thoroughly examined the relationship between higher and secondary education.[3] To the limited extent that historians have considered articulation, they have done so either from the perspective of high schools or from the perspective of universities; they largely have ignored national, regional, and state initiatives that attempted to bring the educational levels together. Edward Krug's seminal work on the shaping of the American high school, first published in the 1960s, explored, more than any other study to date, some of the ways in which national committees and regional associations approached articulation, but he studied the issue mainly from the perspective of the high schools.[4] Studies of higher education similarly are one-sided, concentrating mainly on the university perspective. Harold Wechsler's important study on selective college admission policies, for example, identifies many of the early movements toward articulation but does so primarily from the point of view of higher education.[5]

This book builds on the work of these and other historians to show that the relationship between higher and secondary education evolved in tandem between the two educational levels. It explores the role that both played in restructuring education. It often appeared that the nation's colleges and universities were the dominant players in this relationship, and they did have a lot to lose and gain from a fully articulated system of schools. The secondary schools, however, shaped the debate in profound ways. Articulation ultimately was a process in which both levels played strong, viable roles. In the process, the evolving relationship between higher and secondary education affected what it meant to be a student at the turn of the twentieth century and the education these students received.

## II

To tell this story, the book spans the country from the eastern seaboard and the southern states to the Great Lakes region, with a few stops in the Rocky Mountain West and along the Pacific coast. It analyzes national committees and blue-ribbon reports, significant regional variations in New England and the Middle Atlantic states, the Midwest, and the South, and accreditation programs in Michigan, Wisconsin, and other states. Since national efforts to articulate education primarily included wide-ranging debates and reports by prominent administrators, this book emphasizes regional and state levels where specific programs emerged and where high school teachers and university professors had a stronger voice. It tries to highlight differences among these regions and states while also addressing the similarities that laid the foundation for a unified educational structure. It weaves together the perspectives of higher and secondary education from national, regional, and state levels to emphasize the complex history of American education at the turn of the twentieth century.

This campaign to articulate education has to be explored in the context of social and economic changes in society, and chapter one outlines the larger factors giving rise to the creation of an educational system. During the last years of the nineteenth century, both higher and secondary education were growing and assuming their modern form. At a time when industrialization, scientific advancements, population growth, and vast cities undermined the traditional fabric of American society, the division between the two educational levels was not clear. Throughout much of the nineteenth century, the colleges and secondary schools competed with each other for students, rather than working together to form a hierarchical system of schools. This situation would not last, as the nation struggled to make sense of the events and factors transforming it and as ambitious college presidents capitalized on the demands of this society and of the middle class to promote the value of higher education. A college degree increasingly was a practical credential that promised decent jobs and a comfortable existence, and a growing middle class, concerned about its place in the new industrial order, championed articulation.

Chapter two builds from this context to explore the first sustained effort to eliminate this competition and articulate education. Although educators had been discussing articulation in national meetings and journals, the real effort toward connecting the two levels occurred first at the state level in Michigan. There, James Angell developed an accreditation and inspection program that had professors visiting schools and recommending changes in line with university needs. The university recognized schools doing good work and allowed their students to enter the university without having to take an entrance examination. This program brought the high schools of the state and the university into a closer connection and laid the foundation for similar efforts in other states. It also shifted the admissions process from one that enrolled students based on examination scores to one that admitted future collegians on the basis of a high school diploma. The high schools, as a result, were now responsible for identifying and credentialing the students who would continue their education.

Chapter three looks at the spread of the accreditation program throughout the Midwest, New England, and the South. Each region developed different approaches to accrediting schools, but all three embraced a version of the Michigan initiative. New England, however, also focused on an alternative process for admitting students to college and aligning higher and secondary education. While most colleges in the New England states admitted students through an accreditation program, Harvard and Yale refused to abandon the traditional entrance examination. To reconcile these different approaches and to further develop the relationship between the two educational levels, New England's colleges and secondary schools banded together to form the nation's first regional association. This association, which other regions developed in later years, brought colleges and secondary schools together in a formal organization to discuss the proper relationship between the two. Through such regional associations and the spread of accreditation programs, the colleges and universities solidified their place at the top of the educational system, which allowed them to grow and thrive. But, their actions also strengthened the secondary schools by making such schools the predominant path to higher education. A hierarchical system of education was gradually taking shape.

While New England organized its association of schools and colleges, the nation's secondary schools began to play a role in encouraging the colleges to adapt their admission requirements and degrees to the needs and traditions of the lower schools. The secondary schools felt trapped between two competing demands. Traditionally, the public high schools and private academies—with the exception of some preparatory schools closely aligned with colleges—existed to prepare students for the demands of life. Colleges needed them to educate students for study in the higher branches. Although the secondary schools did not shun this responsibility, they struggled to fulfill both missions with limited resources and support. Caught in this bind, they pushed the universities and colleges to alter their requirements to fit more tightly into the structure of the secondary schools. Chapter four, then, explores specifically the role of the high schools in the work of the colleges and universities.

In the early 1890s, the articulation campaign shifted from state and regional efforts to national approaches. Charles W. Eliot, Harvard's ambitious president and the chair of the Committee of Ten on Secondary School Studies, took center stage, and chapter five considers the work and influence of Eliot's committee. The Committee of Ten—the National Education Association's (NEA) ambitious attempt to reform and standardize education—reflected important trends in articulation, and it laid the foundation for the work of future committees, including the Committee on College Entrance Requirements that finished its work in 1899. While its overall effect was muted, the Committee of Ten affirmed the four-year high school as the standard basis for enrolling in college, and it encouraged greater flexibility within the colleges so that their courses of study embraced the modern subjects that were at the core of a strong high school education.

In 1900, most efforts at articulation shifted back to the regional level. The work of new regional associations in the Midwest, the Middle States, and the South brought forward important advances in the accreditation program and the examination system. It was not until 1905 and the emergence of the Carnegie Foundation for the Advancement of Teaching, however, that a national approach began to have a direct and profound influence on shaping a standardized system of education. The efforts of both the regional associations and the Carnegie Foundation benefited from the work of the Committee of Ten and the Committee on College Entrance Requirements. The sixth and final chapter explores the relationship between these committees and the regional and national efforts prominent in the early twentieth century. The Carnegie Foundation was controversial, but in many ways it represented the culmination of decades of efforts at articulating higher and secondary education in the absence of centralized control.

What started in Michigan as a state effort toward articulating higher and secondary education assumed a national scale by the early 1900s. The progress that had been made was significant, and the landscape of education in 1870 was different from the nature of schools in the first decade of the twentieth century. A seamless system of education had not been perfected, but the country was well on its way toward establishing a strong connection between the two levels. Later years continued to see advances in articulation but these efforts were not as dominant or vibrant as they had been in the earlier years. The foundation had been laid, and later efforts built on the work of Angell, Eliot, and others. A brief epilogue considers the accomplishments of these early educators and what later reformers might be able to learn from their successes and failures.

# III

This study addresses how and why two separate educational institutions with their own histories and traditions developed a relationship with each other, worked out goals and purposes in connection to each other, and shaped an American system of education at the turn of the twentieth century. This creation of an educational system reflected the place of the schools and colleges in the midst of social change and underscored the role of education in the growth of the nation. In the pages that follow, I develop this story and try to capture the successes that education reformers had and the consequences of their actions—from the perspectives both of higher and secondary education and of those who valued these reforms and those who bemoaned what was being sacrificed.

Nicholas Murray Butler, while a professor at Columbia in 1891, said "that one of the most interesting chapters in the history of modern education, when it comes to be written, will be the account of the working out of a system of education, elementary and higher, in the United States, suited to the needs and characteristics of the people."[6] I think he was right, and I hope that *The Standardization of American Schooling* illuminates this interesting chapter in America's educational history.

# Chapter 1

## Changing Expectations for American Education

### I

It was a rather momentous call to action when James McCosh proclaimed that the "grand educational want of America at this present time is a judiciously scattered body of secondary schools." McCosh, president of what would become Princeton University, declared at the 1873 annual meeting of the National Education Association (NEA) that these schools were needed "to carry on our brighter youths from what has been so well commenced in the primary schools, and may be so well completed in the better colleges." A charismatic Scotsman who loved a vibrant debate, McCosh was one of the first educators at the national level to call for educational reform and for a stronger set of secondary schools closely articulated with higher education:

> How are our young men to mount from the lower to the higher platform? Every one has heard of the man who built a fine house, of two stories, each large and commodious, but who neglected to put a stair between. It appears to me that there has been a like mistake committed in most of the states of the Union. We need a set of intermediate schools to enable the abler youths of America to take advantage of the education provided in the colleges.[1]

His call galvanized his fellow delegates. They formed a committee to review McCosh's comments and propose ways to build the missing staircase. The committee reported at the next annual meeting, and, being the deliberative body the NEA then was, continued discussion and formed yet another committee.[2]

McCosh, however, clearly had tapped into a broad sentiment in favor of strengthening the role of the secondary schools in leading up to the colleges and universities. One Wisconsin professor, having little use for staircases, stressed that "the stream of education is dammed between the common school and the college." He was adamant that "this obstruction must be removed, and the only way to remove it is to provide intermediate schools to do the work cheaply which is now but partially done, and that in a costly manner by higher institutions" with their preparatory departments.[3] His colleagues understood that they needed a staircase to the upper floor, or a freely flowing stream—that they had to link the secondary schools to the colleges in a hierarchical educational system—if their institutions were to thrive as colleges or, as many college presidents hoped, become full-fledged universities.

Teachers and administrators in the nation's secondary schools—thousands in existence, notwithstanding McCosh's claim to the contrary—understandably saw things differently. These schools had a broad set of purposes that included more than preparing students for college. The public high schools, specifically, guarded their role as the "people's colleges"—the institutions that historically had prepared most students for the middle class and for the demands of life (as opposed to college)—and they did not look favorably on attempts by the colleges and universities to move them beyond this role or, even more troubling, to take over this role. Nonetheless, the diverse entrance requirements of the nation's colleges and universities challenged the secondary schools—many of which periodically sent some students to college—and they hoped that closer articulation might reduce the multiple demands made by higher education. They had real reasons for supporting articulation, although they did challenge many college policies and initiatives. They were not passive participants dragged along by the colleges and universities.

Throughout the country, then, educators from higher and secondary education, sensing the need for improvement in the haphazard structure of American education in the 1870s and 1880s, embraced specific innovations for bringing about the structure of education that McCosh so colorfully described. Why these educators felt the need to construct this new staircase, or to "articulate" the two levels, provides the basis for understanding the steps they took over the next decades. This chapter explores these reasons first by looking at the larger context of American society and education in the last decades of the nineteenth century. Both were in a state of transition, and external factors—such as a rapidly shifting social and economic context and the growth of the middle class—help to explain the drive toward a hierarchical and efficient educational system. The last part of the chapter looks specifically at some of the institutional factors leading higher and secondary education to reach out to each other. Many colleges hoped to flourish and in some cases to become research universities, and the secondary schools sought relief from the competing demands the colleges placed on them. Both supported articulation because it meant clarifying the purposes and roles of each level, and it meant aligning standards and courses so that students could flow seamlessly from one grade to another. What the public high schools eventually discovered, however, was that their role as the people's colleges was undermined by this emerging system of education, which only increased the tensions between secondary education and the higher branches.

# II

McCosh and his colleague in Wisconsin were not alone in calling for more secondary schools and for a stronger connection between them and the universities and colleges. Others may not have employed vivid, descriptive language, but their appeals underscored that America's school system was in a rudimentary state in the late nineteenth century. The point where higher education began and secondary schools left off remained unclear and ambiguous. The expansion of the college system and the growth of new universities—and their distinction from the colleges—further distorted the educational picture. As long as some high schools, colleges, and undergraduate departments of universities competed for the same students, no real system of education that sent well-prepared graduates of secondary schools into colleges and universities existed. The rapid expansion of secondary schools in the last decades of the nineteenth century only contributed to the challenge of creating a finely articulated school system.

Attendance in some form of secondary school increased in the late nineteenth century, even though the number of students, as a proportion of the total population, remained low. Figures are imprecise, in part because of the difficulty of defining secondary education and identifying secondary school pupils, but by 1876 approximately 185,000 students attended some form of secondary school, including preparatory departments of colleges and universities. This figure represented about 4–5 percent of the nation's fourteen- to eighteen-year-olds.[4] Regional variations were pronounced. New England had higher enrollment patterns than any other region. There, 11 percent of young people attended some form of secondary education, and an impressive 17 percent reportedly enrolled in school in Vermont. The situation was less promising in the Midwest where only 4 percent were in secondary schools; south of the Mason–Dixon line only 2 percent attended school. South Carolina and Arkansas ranked at the bottom, with 0.7 percent of people between fourteen and eighteen years of age in secondary school.[5]

Not many young people were getting a secondary school education in the 1870s, and of those who were enrolled in the secondary schools, most left after one or two years and did not graduate. Few of those who did get their diplomas planned to attend college. Only 45,000 of the students enrolled in secondary schools in 1873 were taking college preparatory courses, and only a handful entered college.[6] The situation improved in later years. In 1881, 225,000 students received some sort of secondary instruction through public high schools, academies and other private schools, and the preparatory departments of normal schools, colleges, universities, and scientific institutes. The number of students in school continued to increase dramatically throughout the 1880s and 1890s. In 1890, 310,000 students (not including some 42,000 in preparatory departments of colleges, universities, and women's colleges) attended 4,500 secondary schools. Fewer than 50,000 of these students, however, were in college preparatory programs.[7]

To further complicate the development of an educational system, many of the 323 institutions ranked as colleges and universities—especially in the Midwest, South, and West—in 1873 were little more than preparatory schools. Although over 52,000 students attended some form of higher education in 1873, 25,000 were preparatory students studying in the sub-Freshman or preparatory departments of colleges and universities. Little had changed by 1885, when nearly 32,000 students enrolled in the collegiate departments of 345 universities and colleges, and 25,000 were in preparatory departments. By 1890, however, attendance had increased significantly. Over 47,000 undergraduates attended 430 colleges and universities, and an additional 12,000 enrolled in the collegiate departments of 167 women's colleges.[8] By the late nineteenth century, few graduated from high school and attended college, but the trend was toward enrollment growth across the board.

In this time of growth and educational change, schools reflected a larger society in the midst of dramatic transformation. What McCosh and other reformers saw around them as they attempted to develop a hierarchical system of schools was a nation dealing with increases in immigration, the shift from a traditional agricultural economy to an industrial one, and dazzling scientific innovations that altered the way people understood the world and interacted with each other. In the North most dramatically, manufacturing advanced swiftly after the Civil War, with new products—including barbed wire, internal combustion engines, and steam turbines—proliferating. Alexander Graham Bell patented the telephone in 1876 and the electric utility emerged in the 1880s. Entrepreneurs such as John D. Rockefeller and Andrew Carnegie powered the growth of new industries and helped to establish centralized control and organization as a means of efficiently producing products. Employing such mechanisms of organization and production, monopolies controlled the steel, oil, and gas industries, among others.

These expanding industries required large infusions of labor—usually unskilled—and towns and cities in both the North and the South swelled as Americans moved from rural areas and immigrants entered the nation. Scientific and technological advances in communications and in railroad and transportation networks facilitated this mass movement from rural America to cities. And while the railroads overbuilt and precipitated panic and depression in the mid-1870s, they brought the continent and its people closer together. What had been a country of small communities loosely connected was giving way to a more complex society made up of cities and vast industries connected by a series of rail lines that crisscrossed the continent.[9]

Old ways of seeing and making sense of the world proved inadequate. As the nation and its cities changed, the demand for services increased. In smaller, tightly knit communities, people reached out to each other and helped meet common needs. In the larger, more impersonal cities that were fast becoming home to more Americans, such approaches no longer sufficed. Cities were complex entities, and they demanded a host of new services and well-trained professionals to administer them, regulate their finances, and generally bring some order to the confusion. They depended on professional administrators who could ensure that basic services—fire, police, and sanitation, for instance—were met. Additionally, an industrial economy, while relying primarily on unskilled laborers, required technicians and managers to coordinate newly emergent and complex production processes, and new methods of finance evolved to underwrite the expansion of industrial businesses, in turn requiring an ever-larger system of bankers and accountants.[10] Accordingly, the number of Americans working in professional positions jumped dramatically between 1870 and 1910 from 342,107 to 1.7 million. College presidents, professors, and teachers climbed from 128,265 in 1870 to over 600,000 four decades later. Architects, chemists, lawyers and judges, surgeons and physicians, technical engineers, and editors all saw similarly impressive gains in the numbers of Americans working in these professions, and these new professionals enjoyed a high standard of living.[11]

Reformers championed the crucial role that education could play in addressing and dealing with the challenges wrought by this transformation. Schools, for example, had the potential to inculcate values of thrift and industry and a belief in private property in the minds of the nation's expanding population of young, poor, and urban children. Wealthier Americans welcomed this role, particularly since they feared that the congregation of the poor and immigrants in slums would increase crime, lawlessness, and political uprisings. Education then held out the promise of bringing some order to the confusing world in which Americans found themselves, but it also was instrumental in shaping a professional workforce to meet the needs of a changing nation. For craftsmen and skilled laborers, small businessmen, and white collar employees, who comprised an expanding and tenacious middle class in the nineteenth century, a high school education promised an avenue into stable, secure jobs for their children, at a time when industrialism and large-scale factory production radically reshaped their existence. A high school diploma—even just a few years of secondary education—represented a credential that holders could leverage in a competitive job market that had fewer opportunities for skilled laborers but needed educated clerks and bookkeepers.[12]

The high school diploma reflected the hopes of middle-class parents that their children would not be consigned to a life of unskilled labor in the behemoth factories of capitalist America. These factories churned out labor-saving devices that middle-class families enjoyed, but these parents did not want their children working on the assembly lines that produced such goods. Middle-class parents thus supported the secondary schools, especially the public high schools, throughout the nineteenth century, and they valued the practical courses that most high schools and academies offered in grammar, history, science, modern languages, and mathematics—subjects they felt correlated particularly well with jobs as editors, merchants, and teachers, and in other commercial pursuits. In theory, students from all backgrounds and socioeconomic classes had access to this education and its promise of future success, but the wealthy frequented private schools or hired tutors and the poorer classes rarely could afford to forego the loss of income by sending their children to high school. As a result, the high schools predominantly benefited the middle class, and enrollments in secondary schools, especially in the urban areas that could support high schools, expanded throughout the nineteenth century (while remaining a small proportion of the overall population). Even smaller towns and rural areas, hoping to capitalize on the benefits of a secondary education, tacked on a few grades to the grammar schools and called them high schools, but these meager institutions could not compete with their more robust, urban counterparts.[13]

Still, by the turn of the twentieth century, as secondary schools prospered and more people held a diploma from such schools, the high school's value as a credential-granting institution began to decline. The demands of an increasingly complex society for professionals—engineers, scientists, professors, and managers, in addition to lawyers and doctors—similarly lowered the worth of a high school degree and its emphasis in earlier decades on preparing students to be teachers and clerks. The connection between the secondary school diploma and a middle-class existence—so vital to the middle class for much of the nineteenth century—was weakening, and this class turned to the colleges to provide the credential that would unlock the doors to a solid professional existence. For this class, a college education promised access to the professions and a higher socioeconomic standing than the high schools could offer, and in the last years of the nineteenth century and the early decades of the twentieth, the middle class came to embrace a college education and the prestige it afforded.[14]

That higher education progressively focused on and embraced the best that scientific research had to offer only enhanced the value of a college education. Scientific research had the power to develop new industries and innovations and to ameliorate social problems. As one educator put it in 1891, a scientifically trained mind—with its ability to observe, classify, and judge—had a quality that "spans great rivers with bridges above the tallest masts, that tunnels mountains, that invents telephones," and "that discovers the causes of disease and removes them." Such a mind possessed a quality "that investigates social phenomena and suggests remedies for existing evils, that purifies the water supplies of great cities," and "that makes the luxuries of modern life possible."[15] Following this scientific impulse, higher education, specifically, established specialized science and engineering departments in the late nineteenth century to train an increasing number of engineers and scientists who would organize and run the nation's expanding transportation and communication networks.[16]

Outside of the actual sciences, professors similarly engaged in scientific research. By the 1890s, the social sciences were burgeoning, and university professors drew upon them in applying their skills to economic and social problems. Through their study and research, economists, sociologists, and political scientists gained an expertise and authority that university administrators promoted as they struggled to build strong universities dedicated to expanding knowledge, developing new technologies, and tempering profound social issues.[17] Even some in the secondary schools felt this scientific impulse and saw it as the means for social progress, but university builders argued that only their institutions—not the high schools or even the colleges—were capable of amassing the necessary resources, money, and laboratories so that faculty and students could employ the best that science and scientific research had to offer as they helped society on its march toward the future. The high schools could lay the foundation, but the universities would take it from there.

By the turn of the twentieth century, a college diploma represented a singular achievement and gave its bearers some of the prestige and status accorded science and higher education and, thus, an advantage when competing for the professional positions that were becoming necessary in an industrial society. Whether or not they entered occupations that specifically required college degrees—and most positions did not—college graduates carried with them an air of authority and leadership that set them apart from other Americans. While higher education rarely provided an education that aligned precisely with professional jobs, a college degree did offer employers a way to quickly identify promising applicants. It signaled that potential employees had a set of skills and abilities that would suit them well in a job. What America's businesses and industries needed were managers, engineers, and other professionals who, without cumbersome supervision and oversight, would embrace the goals of the business or company, remain loyal to it, cooperate with other employees, and supervise lower-level workers. Having progressed through a stratified educational system—one that was becoming even more articulated—college graduates were socialized to hierarchy and order, had learned how to work well within it, and could speak and communicate clearly with their peers. In higher-level positions, employees needed to be able to sort through company manuals and policies, think critically about the work and direction of the company, and conduct their own research as they made independent decisions and solved problems. In other words, they needed some of the expertise central to scientific research, which higher education valued and promoted.[18]

As David K. Brown has argued, a college degree, then, indicated that the degree holders shared a common culture and value system that transcended other character traits, such as religious affiliation, social class, and ethnicity. It was a way for those charged with filling professional ranks to recognize individuals with similar characteristics, skills, and attainments and, thus, reduce the uncertainty involved in hiring large numbers of employees. As a result, college graduates were well-suited to emerge as favored candidates for professional and managerial positions in the nation's reputable businesses and corporations, banks, railroads, and public utilities, in addition to the more traditional fields of medicine, law, teaching, and ministry.[19]

A growing middle class prized this credentialing function, and it took advantage of the access and opportunities that a college education increasingly provided. For this class and for the few students from the lower classes who could afford the time and money involved, a college education promised an advantage in securing professional, middle-class positions and in climbing the socioeconomic ladder. A college diploma was a valuable credential. For a college or university degree to carry such weight, however, it had to be distinct from and in addition to the basic education and diploma provided by the secondary schools, which in the early decades of the twentieth century were rapidly becoming mass institutions. Thus, the campaign to articulate education and create a hierarchical structure took on added importance. That their children should have a solid preparation in secondary and higher education and the occupational status thus provided became a mantra for the middle class, and this class helped drive the campaign to build strong universities and colleges and to connect these institutions with the lower schools.[20]

# III

In this rapidly altering society, education was becoming demonstrably more important, and the number of schools and students in them was rising in response to social changes. But McCosh was right. A strong system of intermediate or secondary schools throughout the country—one that was capable of leading students to college—did not exist, even as progressively more students sought secondary education. Consequently, building the articulated system that reformers wanted and that middle-class parents saw as essential to their children's success was challenging. There were many schools scattered throughout the country claiming to offer secondary instruction, but they did not represent something distinctly or uniformly recognizable as secondary education. Their quality varied widely. "The line of demarcation between elementary and secondary and between secondary and superior instruction is not very distinct, if drawn at all," reported the U.S. commissioner of education, in 1872.[21]

The lack of a distinct model for secondary schools and of a recognizable place in America's loose structure of education challenged McCosh and other reformers. Academies and endowed schools, public high schools, normal schools offering preparatory training, private preparatory schools, and secondary school departments of universities, colleges, and women's schools all claimed to provide some sort of secondary instruction. Once again, regional differences were significant. Three-quarters of New England's secondary students attended academies, endowed schools, and preparatory schools. Public high schools, although a significant presence in Massachusetts, were not readily found in much of New England. Southern students had few opportunities for secondary education in a region still struggling to deal with the aftermath of the Civil War, but private schools there provided some form of secondary education for more affluent students. Only in the Midwest were public high schools a significant presence. In this region, they were growing in numbers and prominence, but even here in the early 1870s, a large proportion of students attended private schools or preparatory departments of colleges and universities.[22]

Many of the private and endowed academies offered excellent courses as preparation for college and for life, as did a number of the urban high schools. Some of these larger high schools and academies even offered courses and programs that exceeded those available in several colleges. But many of the public high schools were secondary in name or intention only. The U.S. commissioner of education bemoaned the state of public high schools in 1873. "While many of these [public high] schools are of a high order of merit and afford excellent training for the colleges and schools of science," he claimed, "it is nevertheless the common observation of experienced educators that a large proportion of the class do not meet present requirements, either in the quality or extent of training, whether the destination of their pupils be the college, the school of science, or business. Many of them are doing the work of the primary school."[23] An editorial in *The Nation* for 1874 agreed with the commissioner and condemned that haphazard growth in which "an infinitude of schools" claimed to offer high quality secondary work, but rarely did so in practice.[24]

While many secondary schools failed to offer a comprehensive program of study in advance of the primary grades, a few optimistically hoped to elevate themselves to a higher rank by calling themselves colleges, as Carleton College in Ohio and Arkansas College did. At best, these schools offered little more than a secondary education.[25] Merely adopting the collegiate moniker or cloaking an institution in collegiate trappings did not make it a college or university, as many "colleges" discovered. These institutions—especially in the Midwest, South, and West—wrestled with the question of how to raise standards, and, in some cases, found it hard to offer more than mere preparatory courses. Milwaukee's S. R. Winchell thought that these "numerous so-called colleges and universities" were "an egregious abortion, neither high school, college, nor university but a kind of overgrown academy or seminary by a misnomer termed colleges." He had no patience for their existence and argued that "it would be an educational boon to this country if some stroke of fate would sweep from existence many of these pretentious, pride-fostering, John Smith incongruities sometimes known as colleges."[26]

Whether Winchell would have included many of the newer state universities in his list or not, he would have found much to condemn in the work they were doing. The University of Kansas, for example, enrolled 77 students in its collegiate program but over 150 in its preparatory department in 1873. The University of Minnesota was hardly different with 232 preparatory pupils and 44 collegiate students. Even Winchell's own state university was not completely safe from slipping into being a "pretentious, pride-fostering" academy. The state university in Madison was slightly more balanced than its counterparts in other states, with 45 collegiate and 35 preparatory students, but the institution had grown dramatically by 1878 and enrolled 281 collegiate students and 120 preparatory pupils. Universities and colleges thus frequently spent significant time, money, and other resources on secondary instruction.[27]

Some universities outside of the Midwest and many smaller colleges throughout the country also enrolled more students in preparatory programs than in advanced courses. Pacific University in Oregon had 100 pupils in its preparatory program in 1873 but only 5 in its collegiate course. Of the 925 students at the College of the City of New York, 592 were in the preparatory program, and 142 of the 195 students at Western University of Pennsylvania were preparatory students. The University of South Carolina enrolled 61 collegiate students and 70 preparatory students. The ratio was more heavily skewed at the University of Nashville, with its 44 collegiate students and 179 preparatory pupils. Smaller colleges similarly found themselves in the preparatory business. Adrian College in Michigan offered collegiate work to 31 students and preparatory work to 113, while Cornell College of Iowa had 75 collegiate students and 250 preparatory students. These colleges and universities accepted students without any solid preparatory training.[28]

Hoping to identify itself clearly as a university, the University of Michigan refused to open a preparatory department. It could do so, in part, because the Ann Arbor high school, where the university was located, acted as a feeder school. Similarly, universities and colleges in New England—a region with a strong tradition of preparatory schools closely aligned with prestigious universities and colleges—were less likely to have preparatory programs than were schools in other regions of the country. Harvard and Yale, as well as Brown and Wesleyan, did not enroll preparatory students. Of the 4,200 students studying in New England colleges and universities in 1878, only 370 were in preparatory departments, but over half of the students in colleges and universities in the Midwest and nearly half of those in the South were taking preparatory courses.[29]

This state of affairs in American higher education led one secondary school administrator to quip that the term "college" included "a class of institutions ranging all the way from a second rate grammar school to a university. Surely, if there is chaos anywhere in our educational system, it is in the field of so-called higher education."[30] He had a point. Higher education comprised a diverse and constantly shifting group of institutions. The categorization of a number of schools offering some sort of higher or advanced education was ambiguous. Educators struggled to classify normal schools, schools for the "superior instruction of women," scientific institutes, commercial schools, and independent professional schools (including law and medicine), which often competed with colleges for students. These institutions, varying widely in quality, offered programs ranging from basic preparatory work to more advanced study. And, confusing the situation even more, some secondary schools, especially in urban areas, offered normal school instruction.

Even something called a college was not easily described. Often differing by region, America's colleges offered students varying educational experiences. Classical colleges, dominated by Harvard from its founding in 1636 and other venerable New England institutions, emphasized classical courses in Latin, Greek, and mathematics, along with some science and rhetoric. These classical colleges historically had trained the young to enter the ministry and to ensure that the colonies and later the states had an adequate supply of well-trained, literate clergy, teachers, and statesmen. This focus on classical preparation—which Yale famously affirmed in its 1828 report on the collegiate curriculum—remained a core part of colleges, even as many colleges gradually enveloped more programs and courses throughout the nineteenth century to meet the needs of a changing society.[31]

While these established colleges dominated in New England, secular state institutions and church schools competed in the South, and religious denominations launched new college building campaigns in the West. These newer colleges spread in response to a growing demand for some sort of education and denominational training. As the transcontinental railroad shuttled people away from the Atlantic coast and the populated regions of the South and Midwest, religious leaders worried that the population would lose its moral footing. Building denominational colleges in remote areas to train clergy and educate the young became popular rallying cries for religious denominations. Town boosters also competed for these colleges and the prestige they could provide, since colleges connoted culture and helped to establish the respectability of a community. But, in the absence of a network of secondary schools, these smaller colleges— usually located in rural areas or along the frontier—had to provide a secondary education. They met a need for education in regions where access was limited, but the education consequently was often rudimentary when compared with the larger, more established liberal arts colleges in New England.[32]

Whether in the western regions of the country, scattered along the eastern seaboard, or rooted in the South, the colleges often were in competition for students with high schools, academies, scientific institutes, professional schools, and normal schools—all of which offered some amount of preparatory work. There was no consistent standard for what constituted a college, and a college degree was not a prerequisite for advanced study in law, medicine, and theology in the last half of the nineteenth century. To complicate the picture, in the decades following the Civil War, many smaller colleges hoped to attract new students by developing nonclassical degree programs, science and engineering courses, women's colleges, teacher training programs, commercial and business courses, and even summer and evening schools, while continuing to promote the traditional classical curriculum. Still, average college enrollment in 1890 was only around one hundred students. Even as they expanded programs and courses, the colleges, especially the small ones, were losing their standing in the educational hierarchy to the emerging research universities—with their graduate and professional programs, libraries, and laboratories, and clear focus on original research and analysis—which in the new century would come to dominate the disparate institutions that comprised higher education.[33]

What separated secondary schools from colleges and universities, even what distinguished colleges from universities and other institutions offering advanced education was not entirely clear at the turn of the twentieth century. The U.S. commissioner of education conceded that "the programmes of study in each are without exact definition."[34] Educators throughout the country asked a common question: Where "should secondary or prescribed education terminate, and where should the special or higher university work begin," as Henry Frieze, the University of Michigan's acting president, put it in 1881.[35] The stairway between the elementary schools and the universities and colleges was incomplete. High schools, preparatory schools, and private academies existed, but they did not always provide the education that the nation's colleges and universities wanted, as McCosh made clear. However, the universities and colleges were not as complete as he thought they were, either. The whole structure needed work. In this context, articulation meant bringing greater order and clarity to the many institutions loosely comprising something called higher education, defining concretely the responsibilities of secondary education, and bringing the two levels into greater alignment.

# IV

Higher education particularly needed to bring order and organization out of this chaos. Secondary schools, too, had strong reasons for supporting articulation, but the urgency for colleges and universities was more pressing. Their ability to survive and prosper depended on the strength of the lower schools. America's universities and colleges yearned to be free from the constraints of preparatory training, and many had ambitions to flourish as centers for original research and advanced study. Their role, as they conceived it, was to prepare tomorrow's leaders in research and the professions, not to drill students in basic subjects.

Some of America's colleges—the state institutions in Michigan and Wisconsin included—and many of the nation's premier schools for higher education, such as Harvard, led by an ambitious Charles W. Eliot, wanted to be full-fledged universities. They wanted to be exciting centers for investigation and discovery, where professors embraced research and tackled pressing issues, where students came to absorb what humanity had learned, and where a few made their own discoveries through original analyses and investigations. These emerging universities championed a place for themselves that was at the pinnacle of discovery and that was the crown not just of the educational system but also of knowledge and learning. The organizational charts that littered presidential offices detailed finely wrought institutions that incorporated teacher training programs, undergraduate colleges, scientific institutes, professional schools, graduate programs, research laboratories, and libraries—all resting on a series of strong preparatory schools.[36]

These universities existed more in the minds of presidents and in detailed plans than they did in reality in the 1870s and 1880s, but over the following decades they came to dominate higher education. As the universities began to thrive, the smaller colleges struggled to keep pace. Usually located in rural areas, which constrained their opportunities to recruit students, most were too small to sustain the larger faculties and advanced programs that increasingly defined universities. Under pressure to find stable funding, most small colleges were not well positioned for change and growth. Some faltered and eventually became academies, while a number of ambitious ones made the transition to university status. Most, however—whether affiliated with universities as undergraduate colleges or not—settled into an intermediate place between the secondary schools and the graduate and professional programs of universities, and they began to carve out a niche for themselves as liberal arts colleges focused on general knowledge and culture. While once they had dominated higher education, colleges were now a smaller part of a larger structure.[37]

To rise to the status of research universities or even to flourish as under-graduate institutions, colleges required secondary schools able to mold students for advanced study and capable of giving students a foundation for the exciting work that would follow, but, as the universities and colleges were coming to understand, not many secondary schools—especially public high schools—existed that could take on such a role. Again, it was Princeton's president who sparked this debate. "The principal difficulty which American colleges have to contend against," warned McCosh in 1873, "lies in the want of preparatory schools in most of the states of the Union, and in the deficient character of the training in many of those academies which propose to fit young men for college." He emphasized his point. "The colleges ought to know that, if they are to live and prosper, they have to encourage the institution of schools fitted to feed them."[38]

Not many of the nation's university presidents needed such warning. The president of the University of Missouri recognized the impossibility of raising university standards and scholarship "without first improving our preparatory schools."[39] Similarly, when James B. Angell was inaugurated as president of the University of Michigan in 1871, he underscored the impor-tance of strengthening the lower schools. "If now we are to lift the grade of university work, we must lift the grade of preparatory work," he told the assembled crowd, "and receive our students only at a more advanced stage of training than they at present reach before entering the Freshman class."[40] Henry Frieze, who was Angell's predecessor as acting president of the University of Michigan in 1870, understood the argument. "If a genuine university is ever to exist, either here or anywhere else in America, it is to be built on a much higher scholarship in the preparatory schools and academies."[41] One Wisconsin professor, echoing the relationship that was evolving between higher and secondary education, recognized the truth in these statements. "While it is true that our higher institutions will deter-mine in a large degree the character of our primary schools," S. H. Carpenter argued, "it is no less true that the success of our higher institutions depends vitally upon a steady supply of students from schools of a lower grade."[42] As these and other proponents of higher education recognized, poorly prepared students with barely a secondary education would never suffice for institu-tions determined to grow and thrive through advanced courses and robust research programs.

These universities needed a growing number of well-prepared, well-trained students to fill their classrooms, lecture halls, and laboratories. For colleges and universities, the staircase rarely connected their institutions with the lower schools, and they often had to open preparatory depart-ments. This decision led to the awkward situation where these institutions enrolled more students in preparatory work than in collegiate study. Such departments did lead students to the higher branches, but they also drained resources and faculty time from what professors considered to be the more important work of higher education, and some leaders of secondary schools, as they grew surer in their own field of work, argued that such departments unfairly competed with them. It was imperative, the colleges and universities decided, that they jettison their preparatory work, but they could do so only if secondary schools expanded in quantity and quality. Preparatory training, they determined, had no place in modern American higher education; it only siphoned resources from these institutions' true work. Nonetheless, in the late nineteenth century, universities and colleges often found it necessary to maintain such departments as feeders to the advanced courses they offered.

A few did not need to rely on such departments, however. Harvard and New England's other prestigious universities had long-established connections with preparatory schools, and they were in a stronger position than the newer public universities in the Midwest and West. Harvard, for one, had built strong ties over decades with a number of preparatory schools, and most of its students came from a handful of these schools, including Exeter, Phillips Andover, and Boston Latin (a tax-supported preparatory school that sent a number of students to Harvard). Tightly aligned with Harvard, these schools ensured that Eliot's institution had a constant supply of students throughout much of the nineteenth century. Harvard in the 1870s was in a fortunate position. No dammed stream existed between Eliot's Harvard and the lower schools.[43]

Harvard would not remain so lucky. In the decades after 1870, Eliot came to recognize that the stream was not running as fast or as wide as Harvard's aspirations demanded. Like many other institutions, Harvard wanted to be more than a well-regarded college that educated the nation's young in the classical and modern languages, mathematics, and a little English and history. Eliot had grander plans for Harvard beyond building a true university with advanced studies and a wide array of courses. He envisioned Harvard as the nation's premier university. To reach this status, Harvard needed more students than its preparatory schools could provide. Never short on ambition and never one to shy away from a challenge, Eliot embarked on a campaign to find the students he needed.[44]

He joined McCosh and a number of others already working to strengthen the secondary schools and their connection to higher education. Angell and Frieze in Michigan and John Bascom, the president of the University of Wisconsin, had struggled with this issue since the beginnings of their administrations in the early 1870s, and they understood something that Eliot initially resisted. The public high schools were quickly becoming the only viable foundation for their universities. Few academies existed in the Midwest, and it proved to be problematic and costly for universities to retain preparatory departments. These universities necessarily turned to the public high schools for students. In doing so, they recognized a momentous shift in the educational landscape. The nation's public high schools, expanding rapidly following the Civil War and propelled by the shift to an industrial, market-based economy, became the dominant form of secondary education in the 1880s. Although southern states continued to favor private academies, elsewhere these academies, once the primary form of secondary education, no longer held that position.[45]

In all regions, with the exception of the South, more students attended the public high schools than the endowed academies and private secondary schools by the late nineteenth century. While 43 percent of students in southern states in 1890 enrolled in public high schools, 83 percent in the Midwest and 73 percent in New England did. Overall, nearly 212,000 secondary students enrolled in public high schools, while 98,000 attended private schools. However, a larger percentage of students from private schools were on the college preparatory track. Only 25,000 public high schools students, or 12 percent of the total enrollment, were preparing for college. In the private schools, 21 percent, or 21,000 students, were taking college preparatory courses. Regional variations provided interesting contrasts. Of the 49,000 students in secondary schools in New England, only 27 percent attended private schools, but 41 percent of the region's students preparing for the classical or scientific course in college were in private schools. In 1873, 83 percent of New England's college preparatory students had been in private schools. Around half of the South's secondary students studied in private schools by 1890, but almost 80 percent of the region's college preparatory students did so. Over 80 percent of college preparatory students in the Midwest, however, enrolled in public schools.[46] In New England and the Midwest, and eventually in the South, the trend was toward the public high schools.

The rise of the public high schools to this position of dominance created special problems for universities and colleges. The high schools had broad public support by the 1880s and most cities and towns had at least one high school, but there was no given assumption that a public high school automatically would—or, indeed, could—meet the growing demand from higher education for well-prepared students. The curriculum of these schools—as well as many of the nation's academies—did not align tightly with the expectations of Eliot's Harvard or even the University of Michigan. Although they were gradually expanding their entrance requirements to include modern subjects and to reduce the heavy expectations in Greek and Latin, the nation's colleges and universities continued to expect college-bound students in the late nineteenth century to have a knowledge of classical languages (Latin certainly, if not always Greek) and of mathematics.

The high schools and many academies, with different historical precedents and traditions, emphasized a more practical curriculum rich in modern languages, English literature, history, and geography. Some added drawing, bookkeeping, commercial arithmetic, and other practical courses to these subjects. The classical languages were a small part of these secondary schools, with the focus on Latin, Greek, and mathematics strongest in the few New England academies whose mission aligned closely with the region's colleges and universities. The larger and better-established high schools rivaled these private schools in courses and opportunities, and some offered strong classical programs. The smaller academies and public high schools, especially in rural areas, however, were hard pressed to offer advanced courses in ancient languages and to meet college entrance requirements, and very few students demanded to study such subjects anyway.[47]

Most high schools did not regularly send students up to the colleges.[48] Milwaukee's high schools, for example, enrolled 2,186 students between 1868 and 1880, but only 16 of these students ever entered college. Milwaukee was not unusual. Throughout Wisconsin, according to one estimate, an average of only 1 student from each school annually entered the university.[49] The high schools were not prepared to do the work of sending students to college. There simply was little demand for them to do so. High Schools evolved out of the lower schools in the early decades of the nineteenth century, and they took their students up from the lower grades and gave them an education designed to prepare them for vocations and jobs. Rather than being schools preparatory to college, they were the capstone of the educational system for many students. They were the people's colleges.[50]

The universities and colleges, their growth constrained by the need for well-prepared students, began to chip away at this mission. As university presidents realized, the public high schools provided the only new source for the students higher education desperately needed. The public high schools may not have been supplying all of the students higher education wanted, but university administrators and professors hoped that the public schools soon would. Throughout much of the late nineteenth century, they accordingly worked to expand the mission of the public high schools.

They developed a two-tiered approach to their campaign to get more students from the high schools into the colleges and universities. They sought to alter the climate of these schools so that college attendance was emphasized and so that universities—not high schools—were promoted as the crown of the educational system. Putting the university at the educational summit and encouraging college attendance would not alone make the universities happy. Their seemingly insatiable demand was not simply for students but for well-prepared students capable of advanced work. To accomplish this goal, they also needed to elevate the standards of the lower schools.

The campaign to place universities at the top of the educational pedestal—with undergraduate colleges nestled beneath them—did not take long to begin. McCosh's initial call to devote more time and resources to building the staircase—or the intermediate schools—between the primary schools and the colleges led to the formation of a committee in 1873 to discuss the ideas he had raised, to survey high schools and colleges throughout the country on ways to bring about a closer connection, and to report the following year. This committee argued forcefully that high school teachers had an obligation to encourage their students to continue their studies. The committee assailed the lack of cooperation between the two educational levels as "utterly disreputable." The problem was not the colleges but the many high school teachers and administrators who "seem to think it degrades them to say that colleges or anything else does anything to complete the education they give; so they are ambitious to have their own schools the top of the system, and resist everything that implies a superior to themselves. In all this they make themselves ridiculous." After landing this blow, the report added that "high school people" were living in a dangerous fantasy world where names substituted for reality. "Calling [the high school] the completion of a course of education does not make it so—names do not change things. All that comes of such feelings is to pamper a blind pride and injure their pupils." Promoting the high schools as the end of education—as the people's colleges—they argued, limited the opportunities students had for finding honorable callings and leading rich, productive lives in a nation that quickly was becoming more industrial, urban, and technologically advanced.[51]

This harsh assessment of the secondary schools marked a substantive issue that the universities and colleges had to struggle with in their search for additional students. A notion of education that ended with high school created an atmosphere that blocked students from ever contemplating further study. As this same report noted, "Young people are extremely apt to absorb notions unconsciously to themselves, and if the high school course is set before them, and talked about among them, as the end and completion of their education, the tide will so set toward having some definite occupation immediately after graduating at the high school." Colleges had little chance of recruiting these students to their ivied halls, and they desperately needed to alter this atmosphere or climate within the high schools. Doing so was absolutely crucial to higher education's drive to encourage more students to attend college.[52]

No matter how imperative the need was to bring more students into college, instilling in high school students the possibility of advanced study would go only so far in meeting higher education's needs. Beyond looking for additional students—and, for some smaller and newer colleges that was the only goal—many universities and colleges wanted students better prepared and ready for advanced classes and studies. These universities hoped to eliminate their preparatory schools and transfer the work of the first year and, for some, much of the second year of college to the nation's academies and growing system of public high schools. Andrew West, a Princeton professor, predicted that such a move would group all of secondary education where it properly belonged, in the secondary schools, and, thus, would help these schools grow and improve. "Once gathered together [secondary instruction] can be dealt with as a unit, and will be far more likely to improve than at present; and as it is really advancing, even under this present disadvantage [of being split between secondary schools and colleges], it will surely advance still faster when the disadvantage is removed."[53]

Bascom, Angell, and other university reformers hoped that they could encourage more advanced work in the lower schools and, thus, in their own institutions by gradually increasing their entrance requirements. They trusted that the secondary schools would readily respond to the challenge of meeting higher entrance standards. As Bascom put it in 1875, "We hope that the intermediate schools—the graded and high schools—will pay special heed to the new terms of admission to the University, and strive to furnish us students well prepared. A portion of them are doing this." Angell, likewise, understood the dependent relationship that his institution had with the lower schools, and he commended these schools for showing "a most praiseworthy desire to push up the scale of their work" which, in turn, was enabling his institution "to begin our University work on a higher plane." Until the secondary schools met these higher expectations, places like Wisconsin and Michigan were confounded in their desire to flourish as universities or even to exist as respectable colleges. Eager to reach such a goal, Bascom and Angell encouraged the high schools to meet the university's entrance requirements even as they continually raised them.[54]

Presidents such as Bascom, Angell, and Eliot spent much of their administrations discussing the proper articulation between higher and secondary education, but they were not always in agreement. Their approaches varied from each other, representing the peculiar nature of their institutions and their status as public or private schools, the public expectations for education, and the characteristics of secondary education in their states and regions. Eliot, for one, ultimately gained a powerful position in American education from which he attempted to formalize and nationalize a restructuring of the goals, purposes, and nature of secondary education, notably through the NEA's 1893 Report of the Committee of Ten on Secondary School Studies. Angell and Bascom had the seemingly more modest goal of effecting a seamless public educational system in their own states through the power of the university. Regardless of the exact nature of their campaign to shape America's educational system, the nation's colleges and universities in the decades to come focused on encouraging more students to enter college and on motivating the lower schools to do a better job of preparing these students for college.

# V

Although they expressed frustration with the expectations of colleges and universities, most secondary school teachers and administrators were not opposed to fitting students for college. The demands of a changing society meant that the high schools had an obligation to prepare ambitious students for higher study. Recognizing this responsibility, few high school educators in the late nineteenth century thought to carve out a separate and specific niche for secondary schools that had no connection with higher education.[55] Rather, they had compelling reasons to support articulation, even though they debated whether the goal should be a complete alignment of courses of study and purposes. The most pressing desire for secondary schools was a reduction in the diversity of college entrance standards. Colleges and universities set their own standards for admission, independent of each other. This variety of entrance requirements challenged secondary schools, since most town and city high schools usually had some students wishing to attend college, and the larger schools struggled with preparing students for a number of different colleges.[56] Uniformity or consistency in entrance standards would make the job of the secondary schools easier, by freeing teachers and students from the constraint of having to meet the different requirements of multiple colleges.

Eliot, in principle, agreed with the need for greater uniformity. New England's secondary schools, he admitted in his 1878–1879 annual report, "are greatly impeded in their development, and distracted in their work by unmeaning and unnecessary diversities in the admission requisitions of the principal New England colleges." He thought the colleges could reduce some of this diversity by adopting "a common standard of examination, in those subjects or parts of subjects which the colleges agree in prescribing." Eliot refused to ask colleges to develop common standards, but he thought it appropriate, where agreement existed, to develop uniform examinations.

If a number of colleges required "four books of Caesar," for example, they should adopt "a common standard of examination" upon those books.[57] Even though Eliot's call fell on receptive ears, wide variation in entrance requirements continued to tax secondary schools throughout the country and the students in them who were preparing for college.

By the last decades of the nineteenth century, colleges and universities had established distinct sets of entrance requirements that fit the courses of study they offered their students. Although these schools generally agreed by 1870 on the importance of Greek, Latin, mathematics, history and geography, and English as college entrance subjects for the traditional classical course—leading to the bachelor of arts degree—they disagreed on what should be required in each of these subjects. Harvard, for example, required Felton's *Greek Reader*—a common reader compiled by Cornelius C. Felton—or the whole of Xenophon's *Anabasis* and the first three books of Homer's *Iliad* in partial preparation for Greek, but Michigan required only three books of the *Anabasis*. Wisconsin expected students to have a knowledge of four books of the *Anabasis* and two books of Homer. Yale required geography but not history, while Harvard specified the history of Greece and Rome. McCosh's Princeton required students to study algebra through quadratic equations, but Columbia examined students in algebra only through simple equations.[58] Schools with a number of students electing to attend more than one of these schools struggled to cover all of the books required in the classical languages, centuries of history, and different mathematical equations. Even in a state like Wisconsin, high school students might choose to attend the state university in Madison, Lawrence College, or one of the surrounding state universities, specifically Angell's University of Michigan. Harvard and other New England colleges were not unlikely options for some students from the West, either. Preparing students for any or all of these schools, with their distinct admission requirements, challenged the high schools.

For those students wishing to enter a different college course—the scientific course leading to the bachelor of science degree being the most likely alternative—preparation in a slightly different set of subjects was required. Harvard's scientific school, for example, enrolled students wishing a more scientific training. The requirements for this school, like other scientific schools and programs throughout the country that awarded the bachelor of science degree, were similar to the classical course but substituted French or German for the Greek and reduced the amount of Latin.[59] These changes and the absence of Greek brought this college course more into line with the public high schools, but this course also diversified the college entrance requirements that secondary schools had to meet.

As more students continued their education in the last decades of the nineteenth century and as the number of universities and colleges they could attend and the degrees they could earn expanded, the secondary schools faced the difficult task of preparing students for the entrance exams at a variety of schools. Unifying this diversity was a pressing concern for them, and articulation meant arguing for consistency within higher education so that entrance requirements reached some level of uniformity. It also meant aligning these entrance requirements with the courses offered in the nation's secondary schools. This lack of alignment was "a very serious evil," a committee of the Massachusetts Association of Classical and High-School Teachers reported in 1877.[60] The secondary schools were willing to reach up to the colleges in rectifying this evil, but they expected these higher institutions to drop down from their pedestal and reach out to the secondary schools. William Collar of the Roxbury Latin School in Boston, however, found higher education unwilling to work with the lower schools. The universities and colleges in this country "are absolutely independent corporations," he said. "They can arbitrary fix, and have so fixed, the conditions of entrance. They contract or extend, raise or lower, their requirements without consulting the schools that feed them and without regard to each other; what they command, the schools that send up students must do." Collar complained that university control of the secondary schools, which could hardly afford to ignore college admission requirements, was "comprehensive and complete."[61] As he pointed out, the nation's colleges and universities did not always show a willingness to compromise with the high schools, and they exacerbated the situation for the secondary schools by periodically tweaking their entrance requirements in the hope that the lowers schools would elevate their standards and align their courses more fully with college expectations.

Cecil Bancroft, principal of Phillips Academy in Andover, Massachusetts, understood the difficulties facing the secondary schools. "Fitting for college, for example, means almost as many different things as there are colleges," he recognized, "and fitting for the science schools means things more mixed and heterogeneous still." Bancroft did not abhor such diversity in collegiate programs. Students had different needs and desires, he claimed. As such, they should have access to scientific programs as well as to the more traditional classical courses. What he disliked was the utter lack of a clear line dividing the secondary schools from higher education. "At present, so far as the secondary instruction leads up to the superior grade and connects with it, there is confusion and misunderstanding which makes the work of the school unnecessarily difficult," he maintained.[62] Not willing to despair, however, he trusted that the two educational levels would succeed in strengthening their connection and better defining the line dividing them. Once they had accomplished this task, the challenge of setting more uniform entrance requirements would be less daunting.

As long as the entrance standards for higher education varied widely, however, secondary school teachers struggled to prepare their students fully, and they complained that they were spending their time preparing the few for college rather than teaching the many who were ending their formal education at the high school level. They worked to reduce the inconsistency in college admission requirements, but they went further and argued that the secondary schools had a broader purpose than sending students to college. In addition to dealing with the lack of uniformity in the requirements for admission to the nation's colleges, secondary schools had to educate the vast majority of students who were not planning to attend college. A classical training in ancient languages, they argued, had little place in the preparation of students who were ending their education with the high school. Although there was some overlap in secondary school courses and college admission requirements—notably in English litera-ture, history, geography, and mathematics—the college emphasis in the last part of the nineteenth century overwhelmingly remained on Greek and Latin. This emphasis on ancient languages, especially Greek, meant that schools were often pressed to choose between emphasizing one language for a few at the expense of modern subjects for the many.[63] The modern subjects, they argued, better met the needs of students as they embarked on careers and jobs, and, indeed, the secondary schools over the following decades pressured higher education to accept more of the modern subjects as admission requirements.

Bringing about greater consistency in admission requirements would free the high schools, teachers hoped, from having to prepare a few stu-dents in different ways for various universities and allow them to focus on what they saw as their proper task—that of educating the majority of their students who were not going to college for the demands of life. But, as American society changed, the number of students seeking to continue their education in college increased, and the colleges and universities argued that they—not the high schools—were better positioned to prepare students for life. A college education, they announced, offered a stronger credential and thus better access to professional positions in society. This changing emphasis began to alter in profound ways the ideas and purposes of education and the relationship between the two educational levels. As more and more students looked to college—although the number remained relatively insignificant compared to the total population—the pressure on secondary schools to meet this need increased. Reducing the inconsistency in admission standards would alleviate some of the pressure on the secondary schools but not eliminate it. As long as the gap continued to exist between college preparatory requirements and the secondary curriculum, secondary school teachers would find it difficult to prepare students for college and for life. Thus, the dynamics were in place for a rather vivid debate between the universities and the high schools over the purpose and nature of education.[64]

# VI

For teachers, principals, superintendents, professors, and university presidents, articulation and linking higher and secondary education raised essential questions about the very nature and purpose of education: What did it mean at the turn of the twentieth century to be an educated person? Did education mean something fundamentally different for students going to college and for those not going beyond high school? Should secondary schools prepare for life or for college, or could preparation for life and college be the same and be met with similar courses? In the decades ahead, educators would repeatedly ask and debate these questions. They would struggle with defining courses of study and the proper work of the secondary schools and of the colleges and universities, and they would wrestle with reconciling the often conflicting goals of higher and secondary education. Their diverging opinions on these questions and the vocal nature of the debate reflected the importance of the discussion and the deep uncertainty about education in a time of rapid social transformation.

What they said and did had a profound influence on the nation's schools and on generations of students. Importantly, the reforms and programs that educators launched in the last decades of the nineteenth century began to shape an educational system where the high schools led up to the colleges and universities. What had once been a relatively flat structure with the high schools and colleges offering comparable courses developed into a stratified and hierarchical system. This system, while initially tolerating broad variations by states and regions, eventually became much more standardized, and students from Michigan to California to Massachusetts came to have educational experiences that were remarkably similar. This creation of a stratified and standardized educational system, in turn, affected who had access to higher education and the benefits of that education, something the middle class valued and demanded.

In developing and shaping this system in the late nineteenth century, neither higher education nor the secondary schools could claim dominance. Higher education often appeared to be dominant—and in many ways universities and colleges were the primary force pushing for articulation, often because they had more to lose and more to gain than did the secondary schools. But the lower schools exerted their own influence on the universities and colleges and shaped them in ways that the universities were not always expecting. Secondary schools proved unwilling to simply accede to the wishes and demands of the colleges and universities. For their part, the universities continued to expect certain standards and levels of academic work from the secondary schools. Together they established a dynamic relationship that evolved over time and that was characterized by debate, negotiation, and accommodation rather than by the dominance of one over the other. Just as neither educational level won a battle for control at the turn of the twentieth century, neither was unaffected by the relationship. As they both discovered, cooperation and compromise proved to be better options.

They had no choice but to work together. No federal agency existed to lead this campaign or to coordinate and organize what was a rather disparate grouping of schools, just as no federal agency gave direction to the transformation of American society. To bring about the system that McCosh and others wanted required a new set of structures and organizations, which educators had to build themselves. McCosh's vivid call for a stairway to reach from the lower to the upper floors helped start the movement in this direction, and others picked up on his call. Rather than this campaign toward stratification and standardization beginning as a national movement, however, it took place in various ways throughout the country, with states and regions developing and implementing their own innovative reforms. The first significant effort came in Michigan, where James Angell, who had just accepted the presidency of the University of Michigan, and Henry Frieze, his predecessor, embarked on a program to substantially alter the relationship between the university and the state's high schools.

# Chapter 2

# Building the University of
# Michigan on a High School
# Foundation

## I

Charles Kendall Adams arrived at the Bay City, Michigan, high school on the morning of June 5, 1884. Adams and his colleague, both professors at the University of Michigan, were there to inspect the school. The visit did not start well. Adams was immediately "impressed by the neglected and unbusiness-like aspect of the Superintendent's room. The dust of ages seemed to have settled down in the room and to have enjoyed already a long repose," he described. Adams maintained this critical manner throughout the inspection. Although laced with humor, his description of the physics classroom condemned the lackluster state of the school. "At one side of the room was a case apparently for apparatus in which a few articles were indistinctly visible," he wrote. "The suspicion was awakened that whatever dust had not succeeded in finding an entrance into the Superintendent's Room had betaken itself to this case." Adams concluded his account with ringing disapproval. "We could not escape the impression that there is pervading the school a general listlessness that is quite incompatible with any high grade of scholarship," but the greatest problem was a rather poor "spirit that seems to have settled down on the pupils, as the dust has settled down in the Room of the Superintendent."[1]

Such was the state of the Bay City high school. But what was Adams, who would become president of Cornell and the University of Wisconsin,

doing at this school? Why did he care about the dust in the superintendent's office, as if he were there to run white-gloved fingers along the
bookshelves and windowsills? What brought Adams to this dusty Michigan
outpost, some one hundred miles north of Ann Arbor and near the shores
of Lake Huron, on a June day?

Adams was there as part of the university's inspection and accreditation
program, an outreach initiative that the university launched in the early
1870s. Along with many of his colleagues at the University of Michigan,
he visited high schools to determine whether they merited being placed on
the university's roster of accredited schools. The stakes were significant for
schools in Bay City and throughout the state. If they earned a spot on the
accredited list, their graduates would gain the right to enter the university
on the strength of their diplomas and without having to take an entrance
examination. The program similarly was crucial to the ambitions of the
university. James B. Angell, who became president in 1871, hoped to propel
the University of Michigan—little more than an undistinguished college
for much of the nineteenth century—into a full-fledged university that
would help Michigan's citizens and others throughout the country adapt
to the challenges of late-nineteenth-century industrial America. To become
such an esteemed university, Angell's institution needed to bring in greater
numbers of talented students capable of taking up advanced subjects. The
university had to establish stronger relationships with the state's secondary
schools, most of which were public high schools, and it had to define and
improve the standards of these lower schools. Toward these ends, Angell
promoted the expertise of his professors—an authority rooted in their professional training and growing interest in research—which, he argued,
imbued them with a responsibility to work on behalf of the secondary
schools and improve education for the good of society. He marshaled this
authority as the basis for the accreditation and inspection program, which
eventually became the model for university-based inspection programs
throughout the country. Adams made his way to the dusty Bay City school
one June day because that was the route he had to take to build the modern
university that he and Angell wanted.

The first part of this chapter outlines Angell's rationale for supporting
the nation's first accreditation program and the motives behind this innovative approach to articulating higher and secondary education in the late
nineteenth century. Following sections explore the logistics of the program
and the effect it had on the university and the secondary schools. Michigan's
accreditation program gave Angell and his professors the opportunity to
interact with the high schools and better understand what the lower schools
were capable of doing. They then used this knowledge in determining how
high and how fast to raise university admission requirements and standards.

Angell further assumed that such a program would increase awareness of the university among high school students and funnel more of them to Ann Arbor. Through this program and by controlling admission standards, Angell began to establish his institution as the head of the state's educational system. The high schools, however, were not unified in their reaction to the growing power and authority of the university. Woven throughout the chapter are their responses. Some, such as the Bay City school, ignored the university's expectations and demands, while others eagerly sought the prestige that university recognition accorded them. Whether they welcomed the university or merely tolerated it, the state's high schools became the gatekeepers to higher education. By admitting students on the basis of a high school diploma from an accredited school, Angell abandoned entrance examinations as the dominant method of admitting students. He essentially shifted the authority for selecting future collegians from his office to the high schools, but even with this added responsibility and prestige, the high schools recognized that their position in the new educational hierarchy was lower than before.

## II

Adopted by the faculty in 1870, the university's innovative program inspected and accredited the first schools a year later. Henry Frieze, who was acting president prior to Angell's arrival in 1871, envisioned using this program to craft a system of education for the state modeled on the German structure, where the gymnasia or special secondary schools fed directly and seamlessly into the universities. Unlike the German system, however, he wanted to provide for universal secondary education. New England, with its tradition of strong preparatory schools and private academies, had something resembling the German gymnasia, but Michigan had only a few academies and a rudimentary system of public high schools. Although struggling to find and establish a strong tradition in the state, the public high schools were becoming the dominant source of secondary education in the Midwest, and thus provided the primary foundation for a state university in need of students. Frieze believed that a system of university inspection of the public high schools would improve these schools and benefit the university. "The effect of this plan of annual examination" of the schools, Frieze declared, "will be to stimulate the schools to a higher grade, and bring them to a more perfect uniformity of preparation, and thus make it possible to elevate the scholarship of the lower classes in the University." Frieze trusted that his proposal would win "for the University a livelier interest on the part of the citizens whose schools are brought into such close connection with the institution," encourage students to attend the university, and gradually improve the lower schools, which, in turn, would allow the university to eliminate much of its elementary work. Frieze rested his hopes for a fully operational, seamless system of education on the inspection program.[2]

When James Angell assumed the presidency in 1871, he fully supported Frieze's initial efforts in building a strong university. He wanted to establish an outstanding institution that would prepare students to tackle challenges and lead the nation as teachers, doctors, engineers, scholars, and active citizens. But when Angell became president, he saw a university that exhibited more potential and possibility than anything else, which he made clear in his inaugural address. "Honorable as has been the history of the University, there is no friend of it who does not wish to see it doing yet higher and larger work," he declared. He further proclaimed,

> The desire of intelligent men throughout the country for a few American universities which shall be to our high schools and even to some of our colleges what the universities in Europe are to the secondary schools of England, the lycées of France, and the gymnasia of Germany is so strong and pervading that it may be regarded as a prediction of the upbuilding [sic] of such institutions of highest grade…Till that end is reached, our opportunities are not seized. Nothing less than that must content us.

To create this shining university, "we must lift the grade of preparatory work," he continued, "and receive our students only at a more advanced stage of training than they at present reach before entering the Freshman class." The task was substantial, since he recommended shifting the first two years of the collegiate course down to the high schools, but, he believed, it was a realistic goal. "The time is not distant when the better and stronger institutions can safely push up their requirements for admission to the standard now reached at the beginning of the Sophomore years, and I am confident that the day is not very remote when they can secure yet higher attainments."[3]

Angell used his inauguration to establish a bold agenda for the university. He proposed a vertical educational system in which his university—with graduate departments, scientific institutes and laboratories, professional schools, libraries and observatories, robust research programs, and under-graduate colleges—rested on top of the independent colleges and high schools of the state. Angell wanted an educational hierarchy, in other words, where he was the chief officer. The reality of education in the late nineteenth century underscored just how ambitious his plans were and how difficult it would be to bring more high school graduates into the university and to align the goals and courses of study between the two educational branches. Angell only had to look around his state to understand the magnitude of the task he had undertaken.

Many of Michigan's "high schools" in the late nineteenth century were small departments added to the lower schools, and they were unsure of their role, especially in preparing students for college. In the decades following the so-called Kalamazoo case—in which the state supreme court ruled in 1874 that public taxation for secondary education was constitutional—the number of schools and students increased. In the early 1880s, 60 schools reported having high school departments for 5,856 students, with another 20 offering a few secondary courses to pupils in the graded schools. Since many of the state's high school students enrolled in a handful of schools, including the Ann Arbor and Detroit high schools, the number of students in the remaining schools was low, and at most, each high school had two–three teachers. By 1890, the educational situation was improving and over 12,000 students enrolled in 153 public high schools. Even with this increase, however, most high schools remained small.[4] These schools complained that they were asked to mold students of varying accomplishments and ambitions both for college and for life. High school principals and teachers were not always sure that they could do both with limited resources, or even what it meant to prepare for college and for life. They protested, with some justification, that a college preparatory focus required a greater emphasis on the classical subjects and a narrower range of courses than many high schools thought was appropriate.

Angell promoted the inspection program as a way to resolve these issues, improve the lower schools, and build an exemplary system of education. He wanted to work out the boundaries between higher and secondary education and concretely identify the work of each level. Clear distinctions needed to be drawn, he thought. Universities were to deal with the higher branches of study, while the secondary schools were to focus on preparing students to take up the more advanced work that the colleges and universities were trying to offer. Angell's goal was to encourage the high schools to focus on preparing more students for college, and to elevate and enhance the work that both levels did for the benefit of society. Echoing Frieze's early ideas, Angell sought to encourage all teachers, from the lowest elementary levels to the university professors, to see themselves as "parts of one united system" working to provide a strong education for all students in the state. During Angell's long tenure as president of the university from 1871 to 1909, he continued to believe in the power of the inspection program to build a thriving educational system. "Perhaps in nothing has the University been more successful to the educational system of the State than in the cultivation of the friendly relation with the schools by the introduction of the diploma system of admission of students," he wrote in his memoirs in 1912.[5] He had to be gratified when other universities and colleges throughout the country quickly adopted it as they too sought to develop a system of education that connected secondary and higher education.

To build the system that Angell championed, Adams and his colleagues—generally working in groups of two—traveled throughout Michigan to visit and inspect high schools. They attended lectures and recitations, evaluated teacher performance, and met with superintendents or principals. They reviewed the published courses of study, compared them with the courses actually offered, and examined their alignment with university requirements. Sometimes they also quizzed students on particular subjects and had them recite specific passages from leading Greek or Latin texts. In practice, the inspections generally tended to be one-day examinations of schools (sometimes they were much more cursory), with the inspectors jotting down a few notes on textbooks used, school discipline, and the number of students hoping to enter the university. The inspectors carefully noted the number and quality of library books and the extent of available laboratory facilities and equipment—usually a dusty old skeleton propped up in the corner. They also paid especial attention to the number of hours in the school day and the length of the recitation periods, with most periods lasting around forty-five minutes and most days around five-and-a-half hours. Ideally, the inspectors after visiting these schools shared their findings with school boards. Occasionally, they also delivered public lectures in the evening on some aspect of education or on their particular field of study.[6]

If the inspectors reported favorably on the schools and the university faculty accepted the inspection reports, schools earned a spot on the university's list of accredited schools. Students from such schools then had the right to enter the university without having to take the formal entrance examination. To maintain accredited status and privileges, schools submitted to a periodic reexamination (generally every one–three years), and students going to the university from these schools had to complete a full preparatory course of study (certified by the superintendent or principal), take final examinations, and receive their high school diplomas. Once admitted to the university, students from accredited schools had to continue to do strong work, since the university retained the right to expel all laggards; graduation from an accredited school mattered little if the new collegians consistently earned poor marks. Angell trusted that this program would make it easier for students to enter the university by removing the obstacle of the examination, an often onerous experience that students had to suffer through during stifling summer months. However, students from nonaccredited schools—especially students from out of state—could still enter the university if they passed the traditional examination.[7]

The benefits to high schools on the accredited list were significant. Because it provided a way for students to enter the university without the hassle of passing an entrance examination, the inspection program freed schools from spending time preparing students to take the exam and reviewing material that might be on it. Schools then could spend this time concentrating on the greater number of students who had no interest in going to college. The accredited schools also effectively gained the right and responsibility to select the students who would enroll in the university. Since the high school diploma, rather than an examination that tested student knowledge, opened the doors to the university for students from accredited schools, the high schools determined which students were strong enough to earn a diploma and a recommendation for advanced study. Ambitious students who hoped to attend college—and the middle-class parents who increasingly supported such education—sought out these accredited schools. Such schools thus could boast of university recognition, which was useful in attracting students and public support. Being recognized by the university was helpful, then, even for schools that rarely sent students to Ann Arbor. Consequently, the number of schools seeking a spot on the university's accredited list quickly grew.

By the end of the first year, six schools were on the list and 50 students entered the university through this program, but eighty-two schools were approved in 1890 to send their students to the university without examination. That year, 164 students entered the university through the inspection program while 131 enrolled after taking the entrance examination. By the mid-1890s, the university inspected sixty–eighty schools annually and rejected an average of two–three per year. Only in appallingly bad cases did the inspectors and the faculty refuse to accredit a school or decide to remove a school from the accredited list.[8] As bad as Adams thought the Bay City school was he nonetheless recommended that the faculty reaccredit it.

The number of schools seeking a spot on the accredited roster kept the faculty inspectors busy. The decision to examine private schools and academies and to extend the inspection program in 1884 to non-Michigan schools only increased the demands placed on the faculty. The number of out-of-state schools seeking accredited status—including schools in New York, California, Illinois, Wisconsin, and Minnesota—quickly grew. The need for students was a constant challenge for the university, and students from other states, especially those from surrounding areas, represented new pools upon which to draw.[9] When the university began to absorb the costs of sending out inspectors in 1891, rather than expecting the high schools to bear these expenses, it may have further driven up requests for faculty visits.[10] To deal with this rapid growth and the growing complexity of the program, the university hired Allen S. Whitney, a former school superintendent, in 1899 to be a full-time inspector, and it empowered university alumni to handle much of the work of inspecting schools in cities far from Ann Arbor and Michigan. Although Whitney and his successors took on much of the inspection work, faculty continued to spend time away from Ann Arbor inspecting schools.[11]

What they found during their visits were schools that varied widely in quality. The inspector of the Detroit High School was suitably impressed: "The teachers whom I saw," he wrote in 1883, "appear to be decidedly clearheaded and apt in teaching; and the pupils were uniformly attentive, interested, [and] possessed of clear comprehension of the subjects taught."[12] Not all inspectors were as positive as this one. Those professors who visited smaller, rural schools often found less to favor. High schools first opened in northern cities in the 1820s, and major urban centers of trade, commerce, and industry such as Detroit often had superior secondary schools compared with fledgling rural high schools. These urban schools, with greater resources, did a better job of meeting university expectations.

While reviewing high schools, the inspectors focused intently on the teaching corps. They paid close attention to the training and qualifications of the teachers and noted whether a school's teachers and administrators had ever attended the University of Michigan. They scrutinized the teachers' abilities to hold the attention of their classes, their mastery of their subjects, and their ability to maintain discipline and draw students into their courses. The inspection committee was pleased with the "superior merit" of the instruction in the Detroit High School, especially that of a Miss Munger. She "seemed to have a thorough command of the subject though many of the questions propounded to the class were rather 'leading,'" the inspector wrote in the early 1880s. "Her method was to go through the exposition of the subject, solicit an answer at every point where an answer was *obvious*. The exercise was partly a lecture and partly a recitation—but more especially the former; and as such was highly creditable to the teacher."[13] At the Big Rapids High School in central Michigan, the inspector singled out Miss Bartram, who was doing some "of the best if not the best work in U.S. Hist[ory] I have ever seen in High School work. Miss Bartram is a natural teacher. Next year [she] will probably have Algebra and Geometry and is sure to succeed."[14]

Not all teachers received such praise. Adams, who was much less biting but no less direct than he was with the Bay City school, noted that the history teaching in the small Alpena High School in northern Michigan "was not so good. Indeed that of Miss Woodward was positively poor, and a change for the better should be made."[15] One Lansing High School teacher lacked adequate mastery of some of his subjects and he had no real "ability to present [those] subjects in a practical and useful way," the inspector claimed. "His harsh and abrupt manner is at present a disadvantage to him."[16] It was not enough, however, for teachers to have command of the subject matter and a natural aptitude for teaching; they needed to possess a masterful ability to control a roomful of students and maintain discipline at all times. As John Dewey, who himself was not a particularly strong disciplinarian, put it in an 1892 inspection report, "Miss Root's teaching [in the Corunna High School] would be good enough, but she has charge of a room and she is not up to the discipline and teaching at the same time." Miss Root suffered the fatal flaw of letting her room get away from her.[17] Hindered by these problems, these schools nonetheless earned places on the accredited list.

The Corunna High School's lack of discipline did not prevent it from being accredited. Its poor reference library and limited laboratory equipment also were not serious impediments to accreditation. Even the continual turnover of teachers and administrators was not serious enough to derail the school's hopes for a spot on the accredited roster. The university inspectors did point out one issue that had the potential to push the school from the list. Between 1888 and 1892, the high school enrolled fifty–sixty students but employed only two teachers, one of whom served as the superintendent. The inspectors repeatedly pushed the school to hire another teacher, which the school failed to do until the 1892–1893 school year. Dewey even met with school board members and made it clear that the school was "altogether too large for two teachers."[18] The frequent demands from the university for additional teachers had little effect on a school with an unstable staff that shifted from year to year, but each year the school continued to hold its place on the university's roster of accredited schools.[19]

This focus on teachers and order was stronger than the focus on students. The inspectors were more concerned with evaluating the teachers than they were with examining what the students actually learned in the classroom. They rarely tested students individually. The assumption seemed to be that students would learn if they were in classrooms with well-educated, high-quality teachers who knew their subjects and could control a potentially rowdy group of students. Frieze initially hoped that the inspectors would focus more carefully on the work of the students, and he envisioned classes preparing written answers to questions devised and evaluated by university professors.[20] Angell, like many of his counterparts, supported the emphasis on teachers, and he argued that such a close focus on teaching ability was key to improving the school. Since Angell essentially was ceding enrollment decisions to the high schools with this program, he had to ensure that good teachers were preparing students for advanced work. The inspection committee "aims not so much to determine the attainments of individual scholars," he claimed, "as to examine the methods of instruction practiced, to judge of the qualifications of the teachers, to observe what is the outfit of means of illustrating teaching, to ascertain how heartily the school is sustained by the public, and, in general, to judge of the organization of the school, the scope and worth of its work, and its probable prospects." Where teachers were found wanting, he said in 1887, "the committee can in a friendly and confidential way call the attention of teachers to errors in method or manner, and suggest remedies." In the case of a "hopelessly inefficient" teacher, Angell believed that the inspectors had "the very delicate duty of directing the attention of the board," thereby suggesting that the board replace the teacher with a more qualified applicant.[21]

In reality, the inspectors always had to balance the poor nature of a teacher against the possibility of finding a better one. In a state with a limited supply of teachers and a growing high school population, a school might easily end up with an even worse teacher.[22] Often the instructors—both the good ones and the poor ones—were graduates of the high schools where they were teaching, and, likely, there was no real desire within the school or the community to replace them. Only in extreme cases of poor teaching, then, did the inspectors publicly advocate the firing of a teacher.

By the 1880s, Angell had come to recognize that one of the most efficient ways to ensure strong high schools and to encourage students to attend the university was to fill the state's teaching posts with well-trained college graduates, preferably those with degrees from his university. He also understood that his institution needed to take on teacher preparation in a formal way. No longer content with sending out graduates trained solely in the disciplines, the university began offering courses, in the late nineteenth century, in pedagogy and the psychology of teaching.[23]

Until the time when the university could fill most teaching posts with its own graduates, it often put up with weak high school teachers, and the inspectors were reluctant to suggest dismissing a teacher. They also hesitated in offering concrete suggestions for improvement, even though the president wanted the inspectors to do just that. The inspectors knew what they wanted from the teachers and could determine fairly quickly whether a teacher was successful, but they never enunciated the qualities that made strong teachers superb instructors. They could point out weaknesses, but they often were unwilling or unable to help teachers improve by suggesting new techniques or ideas. Many of the early inspectors had never been trained in the science or art of teaching, and they may have been poor university teachers themselves. Dewey was the rare professor who studied education and made it part of his life's work, but he was hardly a scintillating lecturer, for example. These professors knew whether teachers understood the fundamental principles of Latin or botany, and they could sense whether a classroom of normally boisterous students was learning anything. When it came to offering suggestions to struggling teachers, many of these early inspectors had little authority or expertise in teaching, aside from their own classroom instruction and experience, to guide them. This expertise could take them only so far.

These interactions between the inspectors, who were predominantly white males, and the high school teachers, many of whom were white females, emphasized larger trends in the feminization of teaching and the professionalization of administration along masculine lines. In their professional positions as administrators, men supervised a predominantly female teaching force and worked to ensure that what happened within the classroom adhered closely to what this growing group of professional educators thought was essential. As historians have noted, professional educators—including university presidents and professors, as well as college-trained high school administrators—joined forces in the late nineteenth century to control education and the schools. They wanted to ensure that their ideas for education, informed by the best business practices and their own growing research and professional training, supplanted the ideas and control of lay boards, many of which served local expectations. The changing needs of society, these professional educators claimed, dictated the importance of a unified approach to education. The inspectors that Angell sent into the schools fit within this trend toward the professional control of education. As such, they represented another level of male authority over the work that increasingly female high school teachers did. At the same time that it launched this inspection program, the University of Michigan also began admitting women to its campus. Pressed by the need for students, Michigan's state university opened its doors to and filled its classrooms with female students. Women were coming to have a significant presence in education, both as students and as teachers, even as men, in particular through the university's inspection program, sought to establish a firm grip over what went on in those classrooms and what teachers taught and did.[24]

The inspectors placed much of the emphasis for a good school on these male principals and superintendents, rather than on the female classroom teachers. The situation in Eaton Rapids, a town in southern Michigan, underscored the importance that the university placed on a strong administrator able to guide and direct a school. The ill-fated superintendent of this school was struck by "an attack of brain fever," which, nonetheless, did not stop the inspector from commenting that his tenure "has not been in all respects crowned with success." His difficulty, in this his first appointment as a superintendent, lay in energizing his teachers to accept his vision for the school. "In some indefinable fashion, there seemed to be a lack of harmonious understanding between him and the pupils and the teachers, attributable doubtless in some measure to his sickness." Nonetheless, the university accredited the school, believing that such approval would motivate the school to do better. The following year, with a new superintendent at the helm, the school again earned university accreditation.[25]

The unfortunate situation in Eaton Rapids was the opposite of the efficient, well-managed school in Lansing, where the excellent superintendent was "a growing man in this field" with "kindly feeling, enthusiasm, and devotion to his work." He led his teachers, who were loyal to him, in building a strong high school dedicated to the goal of preparing students well for the university and sending them up to higher study. The Eaton Rapids school, with its superintendent struck down by "brain fever," could not compare to the Lansing administrator and his more professional institution.[26]

As with their focus on teachers, Angell and his inspectors paid particular attention to the training and college education of school administrators. It was not always the case that inspectors criticized administrators solely because they lacked a degree from the University of Michigan, but it was not unusual for the visiting professors to be more receptive to administrators they had trained. An interlocking directorate of sorts between Ann Arbor professors and local school leaders was in the making. The superintendent in Eaton Rapids possessed only a limited Normal school training, while the principal in Lansing at least had spent time studying in Ann Arbor. The professionalization of teaching and of those supervising and directing schools was a critical issue for the university's inspectors, and Angell quickly realized that preparing graduates to take the helm of the state's high schools and to teach in its classrooms would create stronger relations between the university and the schools. Having sympathetic graduates in positions of responsibility where they would guide teachers and students, encourage school boards to implement stronger courses of study, and point students in the direction of the university enhanced the stature of Angell's institution and the preparation of entering students.

The University refined, altered, and expanded the inspection program over the course of the late nineteenth century, but it continued to insist on personal visits to the schools. These direct interactions with the schools were the crux of the program, and justified for Angell, Frieze, and others the heightened responsibility entrusted to the high schools. After all, the university's innovative program undermined the examination system and gave much of the authority for selecting Michigan's undergraduates to the lower schools. Angell shifted the process of admitting students from one that tested student knowledge to one that relied on the high school diploma as a credential. In this new system, he trusted that the high schools would meet university expectations when selecting students for advanced study. Once he transferred admission from the university to the high schools, the inspection was the only way to test the quality of the schools and, thus, of the students. Not surprisingly, then, the plan aroused opposition.[27]

Charles W. Eliot, the long-serving president of Harvard, was the most visible opponent of the system. Since the inspection was the only "check" on the privilege granted the high schools to send students to the university without examination, Eliot attacked these regular visits as insufficient and "ineffective."[28] For Eliot, the inspection system did not warrant the transfer of responsibility for admissions to the lower schools. The Michigan program, he asserted in 1878, was nothing more than "a lax, equivocal method of visiting schools."[29] Eliot cautioned that too many changes in a school and too many disruptions in the teaching staff occurred in schools for periodic inspections to be an effective and consistent barometer of a school's standing. He further questioned whether any professor had the ability to adequately and competently examine and assess every aspect of a school's character. It is clear, Eliot asserted, "that there is not a single member of the Harvard Faculty who would feel himself competent, without a good deal of special preparation, to examine a well-organized secondary school in all its departments."[30] He doubted that one or two university professors trained in specialized disciplines possessed enough knowledge of secondary schools and their challenges or could spend a few hours in the schools and then competently pass judgment on them.[31] Eliot also questioned Michigan's objectivity in the inspections. He suspected that the university's need for students undermined its impartiality and led it to accredit schools simply for the students that might come to the university.[32]

The inspection program survived Eliot's attacks, but his criticisms had not been without merit. The University of Michigan had no preparatory department and only a few high schools were capable of preparing students for entrance to the university. The need for students was a compelling factor in the emergence of Michigan's inspection program, and indeed Michigan did liberally accredit schools.[33] By doing so, it weakened one of its potentially most effective tools—university recognition—for directly improving the lower schools. Furthermore, the university hoped to bring students to Ann Arbor from more schools and to open up educational opportunities to students throughout the state. Most of its accredited students, however, entered the university from only a handful of strong, well-run schools.

The weaker schools on the accredited roster were balanced by exemplary schools that received university recognition from the beginning. The high schools in Ann Arbor, Detroit, Flint, Grand Rapids, Jackson, and Pontiac, for example, all enjoyed recognition as solid schools preparing students for the university, and were all on the university's accredited list. These schools were in larger, urban areas with more resources than their rural counterparts enjoyed. The Ann Arbor high school acted essentially as a preparatory wing of the university, and students from throughout the state attended Ann Arbor in preparation for enrolling in the university. Between 1871 and 1880, over half of the students entering the university through the inspection program came from the Ann Arbor high school. Over 80 percent of all students entering the university through this program came from these six schools.[34] These schools earned a spot on the accredited roster for their rigorous preparatory courses, good teaching, excellent discipline, and strong libraries and scientific equipment, but they likely would have had these characteristics even in the absence of the university inspection program. Before these schools even began inviting professors to visit them, they had developed into model institutions, as far as the state was concerned. Rather than molding them as preparatory schools, the inspections only confirmed their exemplary status.[35] The schools added to the accreditation list in later years never rivaled Ann Arbor or Pontiac or Grand Rapids in the number of students enrolling in the university, and most of the smaller schools on the list sent at most two or three students to the university; some years they failed to send any.[36] University professors spent a significant amount of time and effort visiting and inspecting the state's high schools when most of the students entering through the inspection program came from only a handful of institutions.

Moreover, the overall level of scholarship of the diploma students— those entering on the strength of their diplomas through the inspection program—was generally equal to that of the examination students. In 1880, in one of the only studies to analyze the program's effect on student achievement, the faculty compared the diploma and examination students by looking at the number of exams passed by both groups during the first year of college. They compiled these statistics for all students from the beginning of the program in 1871. They found that the diploma students passed slightly more exams than did the other students, although in some years the students admitted through entrance examinations did better or only slightly worse than the diploma students. The diploma students also were somewhat less likely to drop out of college than were the examination students. The message was clear, however: the inspection program had not dramatically increased the level of scholarship of the diploma students in comparison to the examination students. The record may have been "in favor of the students admitted on diploma" but only marginally. The close relationship that the university had developed carefully over time, and with significant investment from university professors, had not markedly increased the quality of the lower schools or elevated the scholarship of the diploma students, as measured by exams in the first year, above that of the exam pupils.[37]

It also was not at all clear that this program brought more students into the university, although Angell at times certainly suggested that it did. To build a strong university, Angell needed public support for education and good high schools to prepare students well for higher education. As was the case with other ambitious universities throughout the country, Michigan needed more students willing to—not just prepared to—pass through its gates and take up advanced study. Since it had no preparatory department and the state lacked a tradition of private academies, the University of Michigan depended on students from the public high schools.[38] Angell and a growing chorus of educators complained that these schools acted as if they were the end of education when they should have been encouraging students to enter the university. A report from the National Education Association (NEA) in 1882 concluded that "the high school is sometimes, perhaps often, conducted as if it were the last stage of the educational process." Such a position, the committee complained, stopped students from considering the university. "From the course of study, from the general animus of the school, and perhaps from the direct influence of the teacher, the pupil receives the impression that there is nothing more to be learned." As a result, the committee claimed, "his desire to secure a complete general education is checked or extinguished," and the final stage of education "virtually ignored. The high school is made the terminus."[39]

Angell's grand plans for education in Michigan championed a new course of action where students regularly and routinely passed from the halls of the lower schools and through to the university's quadrangles. Rather than sustaining a haphazard process where some students might decide to prepare for college, Angell envisioned a formalized progression where the transition from high school to college became a regular, institutionalized part of a student's education. By the program's second decade, Angell claimed that the inspection visits indeed had encouraged more students to consider university study. "To many a school-child, living at a distance from the university, the institution seems something remote, inaccessible. He hardly dares cherish the hope of ever reaching its walls," he said in 1887. "But the arrival in the school of a university teacher, the opportunity to confer with him, his friendly word of encouragement to those who aspire to a liberal education, determine not a few students of character and intelligence to reach the highest grade of scholarship in the school, and to push on to collegiate or university courses." The growing ambition of these students to continue college study, Angell believed, then rippled throughout the school and encouraged even more students to attend the university. "The university is thus brought within sight of all the school children, and distinctly invites them all to come to its open gates."[40] Angell unlocked the doors of his university, and he argued that the inspection program encouraged more students to enjoy unrivalled academic opportunities in Ann Arbor.

Superintendents of two prominent schools on the accredited list agreed that the program built interest in the university among students and led more to consider continuing their education in college. The head of the Peoria, Illinois, school lauded Michigan's plan for landing "the graduates of the high school, not *before* the doors of the college, but *within* those doors." The clear result, he claimed in 1890, was an increase in the number of students wanting to attend the university. Had other colleges embraced the program, he concluded, "then the college contingent found in each class in the high school would be much greater than now."[41] Years before, the superintendent in Pontiac, Michigan—a school that easily earned a place on the accredited list—made a similar argument. In his school, the visits by the university professors excited the students, which led to conversations about the university and its courses. The result, J. C. Jones declared in 1875, was that the inspection program determined "many more on a college course than under the old system. This increased amount of talk is one of the greatest benefits to the school, for it brings the University within the pupil's vision and constantly augments his desire to enter its walls." Over the first few years that Pontiac had the privilege of sending students to the university without examination, the number of students studying Latin—a barometer for interest in college study—shot up appreciably from nine in 1872 to thirty in 1874, even though the number of students in the high school remained constant.[42]

Not only did it lead more to college, Jones believed, it built stronger public support for education generally, which accrued to the benefit of the university. In his school, the superintendent declared, "parents manifest more interest and greater pride in the school and its success" as a result of the program. "They get into closer sympathy with the school, come to understand the character of the work it is doing, and become much more earnest supporters of it."[43] A Michigan professor concluded in 1874 that the inspection program was succeeding in this goal. "As a means of awakening an interest in higher education on the part of the High Schools and the communities in which they are situated, the plan is doing all that its most sanguine friend anticipated." He claimed that even the most doubtful superintendents and principals had come to see the value of the program and "now testify that it works admirably in its influence upon their schools, and on the public sentiment with reference to liberal culture."[44] Since most of this public support and the students in the secondary schools came from the middle class, the university's inspection program and its relationship to secondary education was cementing important bonds with the backbone of bourgeois society.

Pontiac's connection with the university may have been furthering interest among an increasing number of students in the university, but in the inspection program's initial year, only about one-fourth of the entering class came from the accredited schools. The numbers increased considerably in later years. By 1873, nearly 50 percent of the students sought admission based on their diplomas, while in the following year, 60 percent of the new students entered from the accredited schools. In the first ten years of the program, just under 50 percent of all students entering the university came through the diploma program.[45] In this way, the innovation that Frieze developed and Angell championed did succeed in funneling students into the university.

Some university officials, however, questioned whether the program actually encouraged these students to enter the university. An 1883 committee of university faculty that included Charles Kendall Adams doubted the efficacy of the program in bringing more students to the university. "That it has increased the number of our students we see no reason for believing," the committee concluded. This committee did think that expanding the system to students from out-of-state schools would expose these students to the superior advantages of a University of Michigan education and would encourage more of them to travel to Ann Arbor for advanced study. Where students were unfamiliar with the university, the inspection program could be an advantage, the committee believed, but it doubted that the program dramatically had increased the number of Michigan students who would not have entered the university had they been required to take the traditional entrance examination.[46] Michigan's program made it easier for students to attend the university, which strengthened its connection to the lower schools, and it succeeded in encouraging some students to think about a college course of study, but it is not clear that the program brought more students into the university's lecture halls than would have come anyway.[47]

# III

Even with a mixed record in raising standards and increasing enrollment, however, the program was crucial to Angell and his university, and he supported it throughout his long tenure at Michigan. Angell hoped that the inspection program and the prize of university recognition would elevate high school standards, but it turned out that the best way to boost the quality of the state's high schools was by increasing the university's entrance requirements. The inspection program proved to be an integral part of strengthening these requirements. By bringing the university into closer connection with the schools, the program helped the university determine what the schools could do, what the public expected from the schools, and how fast and to what extent the university could increase its requirements. Only a rudimentary educational system existed in Michigan when Angell first climbed the steps to his office in Ann Arbor in 1871, but the inspection program and higher entrance requirements were helping to solidify the university as the crown of the state's system of schools.

Angell regularly consulted with the high schools before taking any steps that affected them, and the inspection program played a crucial role in the university's attempts to figure out what the high schools were capable of doing. "The visit of the university committee to the school," he wrote in 1887, "is of service by guarding the university faculty against the peril of asking the school to do more than it can wisely undertake." He recognized that professors had the tendency to expect the lower schools to meet the advances in entrance requirements "more rapidly than the schools can lift the range of their instruction." When this happened, the gap between the two educational levels widened, and James McCosh's staircase connecting the university to the lower schools became steeper and narrower. Angell diligently sought to avoid such a development.[48]

He also acknowledged a challenge that McCosh, Eliot, and presidents of other eastern universities did not have, at least to the extent that Angell was confronted with it. The University of Michigan, a public institution dependent on the goodwill of Michigan's citizens, relied on another set of public institutions for students. Angell knew that the high schools, rooted in a tradition of local control, could not dramatically alter their mission overnight to become college preparatory institutions. Parents and community members expected the high schools to prepare students in the modern subjects for middle-class positions. Although middle-class attitudes toward college preparatory work shifted favorably in the last years of the nineteenth century, especially as the colleges gradually focused more fully on the modern subjects this socioeconomic class valued, Angell had to move slowly in his campaign to build a strong university on a high school foundation. He refused to back down, however, from his desire to align courses of study and bring more students to the university fully prepared to take up advanced study.

Caught between the wishes of the public and the requirements of the university, high schools had to navigate carefully, which Angell understood. "While the school teachers are always as desirous as the college teachers of elevating the grade of their work," he claimed, "they often find that the tax-payers are not ready to support them in this effort." Angell and other wise university presidents recognized this dilemma and turned an ear to public expectations, and here the inspections were advantageous. "The visit of the committee of the faculty to the school," Angell asserted, "enables them to study the conditions under which the school has to work, to see exactly how fast and how far the school can go, and what limits the university must set to its demands on the school." Only after securing "harmonious co-operation" with the schools, in large measure through the visits of the inspection program, could the university gradually alter its admission standards.[49]

Armed with knowledge from these inspection visits, the university, starting in the early 1870s, began to require the high schools to offer more courses and to prepare college-bound students in more subjects. These additions to the admission requirements, Angell stated, were "in accordance with the general plan to raise the grade of our work as rapidly as the preparatory schools can raise theirs." Often these changes reflected the "modern" courses that most of the high schools had been teaching. Angell was raising standards but he also was fitting the new admission requirements to what the high schools were offering. As part of these early changes, the Latin-Scientific course replaced Greek with French and by 1872 required a full year of French. The following year, the Scientific course introduced "elements of Natural Philosophy, Botany, Geology and Zoology."[50] In subsequent years, the university tweaked its admission requirements but held off making further significant changes until 1889. As science grew in popularity and contributed to innovations in American society, Angell and his faculty again asked the high schools to change their courses of study and prepare students more fully in science. Strengthening the scientific preparation of students entering the university was an important change in its admission requirements, and the change allowed the university to drop some of its teaching in the elementary sciences.[51]

Asking for better preparation in the sciences was crucial to Angell's attempts to further the standing of his institution as a university. He wanted the high schools to focus more concretely on the principles of scientific research and study. Often Angell wanted students to work more with scientific apparatus in fully equipped laboratories where they could develop strong skills in observation and analysis, and in making judgments and drawing conclusions. The inspectors in their reports continually noted the poor state of laboratory facilities in many schools, and they recommended time and again that the state's high schools invest more money in building and equipping scientific laboratories. For an institution hoping to build a reputation for outstanding scholarship and research, bringing in students prepared for advanced work was essential. Moreover, such research skills were crucial in all subjects. Inspectors pushed schools to fill their libraries with outstanding reference books that students in history and in civics, as well as other subjects, could use as they completed their own research projects.

The university clearly benefited from changes in the scientific curriculum, but Angell stressed that he took this step—one that he thought aided the high schools, as well—only after careful consultation with the lower schools. "We do not desire to make frequent changes in our requirements for admission," he stated in 1889. "But these now announced have been under consideration for some years and have been made only after very full consultation with a large number of superintendents, principals and other teachers in our high schools." Angell was careful never to appear to be dictating to the lower schools. "Our fixed purpose is to work in the most harmonious cooperation with our schools and never to make any demands of them which they cannot fairly meet."[52] Instead he emphasized his attempts to work with them in determining what they could do to meet the needs of the university and then thanked them for their willingness to meet the new requirements as part of the march toward progress.

Some of the high schools were willing to adjust to the university's shifting entrance requirements in science and other subjects. After the changes made in the early 1870s, the president of the school board in Detroit reported that the high school course needed revision and that the "whole question of the course of study ought to receive the prompt and careful consideration of the incoming board."[53] The high school in Houghton, with the approval of the school board, similarly altered its courses to meet the new requirements.[54] In his annual report to the superintendent of public instruction, the Owosso superintendent reported that his "college preparatory course has been extended so as to meet the advanced requirement for admission to the University."[55] At the Vicksburg Graded School, the school board worked to align the school's courses with university requirements and requested that Angell send an inspection team to evaluate the high school and accredit it. In the event that the university might find the school unacceptable—which it did not—the school board asked that the university recommend any necessary changes so that it could earn a spot on the accredited list.[56] Indeed, Angell worried that some smaller schools were being too zealous in their drive to elevate the standing of their schools in relation to the university. "So ambitious are the schools to enlarge their work, that the university has found it necessary in many cases to advise the schools in the smaller towns not to undertake so great a variety of work as they have planned, but to do thorough work of more limited range."[57] Some of Michigan's schools were rising to the challenge set by Angell and his university.

These schools trumpeted their relations with the university in their annual circulars and announcements of courses of study. As they sought to increase attendance and even to bring in fee-paying students from other districts, many of the state's high schools emphasized that their courses aligned completely with university expectations. Having a place on the university's accredited list gave these schools further leverage in their attempts to recruit their own student bodies. The superintendent in Pontiac highlighted his school's recently garnered status as a university-accredited school. "This connection of the High School with the University is most salutary in its effect upon the school increasing the interest in and a desire for a completer education," he told his board. "The effectiveness of this High School will depend hereafter greatly upon this very relation."[58] The announcement of the Union School, a small but good school in Vassar, was typical: "The Courses of Study have been arranged with reference to the amount required to enter the University of Michigan," it announced. "Students completing a course of study will be admitted to a corresponding course in the State University, without examination."[59] The Vicksburg school, even though it had not yet received university recognition, pointed out that it had applied for accredited status and that its courses prepared students for corresponding university courses. "The courses of study have been so arranged that graduates may enter the English or the Scientific Course at the State University," it proudly proclaimed, and "as soon as arrangements can be made, they will be admitted *without farther examination*."[60]

Not all schools, however, willingly embraced the shifting demands made by the university. The public high schools in Ann Arbor, Detroit, and Grand Rapids that had long prepared students for the university had little choice but to adopt the new requirements, but these schools also were in a position to exert an influence on the university. Ann Arbor High, acting in its role as a preparatory school to the university, challenged some of the new requirements, not because they were too high, but because they conflicted with the courses they thought their students needed. As the inspection reports showed, it was not unusual for other schools to resist or challenge the university and its expectations. They neglected to hire teachers or to improve their offerings in certain subjects. For all the school boards and superintendents who supported the university, others frowned on the university and pushed their own ideas for the schools. On paper many of the courses of study in the state's high schools aligned with the university's expectations, but the inspectors sought to determine whether the courses as taught met university needs. They found problems throughout the state in the quality and education of teachers, the way teachers taught and what they focused on in their courses. Efforts to raise standards through increased admission requirements and periodic inspections did not always translate into improvements in the classroom, even if the courses on paper changed in line with university courses. Thus, the inspections not only helped the university comprehend what the schools were doing and capable of doing, they also helped Angell and his faculty determine how well the schools were implementing the new requirements. By no means did they find uniformity throughout the state.

As Ann Arbor's challenge to the university suggested, there existed a tension between the work of the high schools and the needs of the university. A number of questions were at the crux of the matter: Were the high schools preparatory institutions that existed solely to feed students to the university, or did they have their own, independent function in preparing students to enter the workforce? Angell saw the high schools as preparing students for the university and for life, and he believed it was possible to do both with a strong academic curriculum. The high schools, conversely, valued their role in sending students out into society with some useful skills—whether that meant commercial courses in bookkeeping or a more complete preparation in the modern subjects of literature, history, and science than they would get in a college preparatory course. For many students, this high school education would be the only preparation they would get. High schools were not unwilling to prepare students for college, but they had to find a way to balance this need with what they saw as the needs of the larger body of students not going to college. For many high school teachers and administrators, the high school was the people's college, and existed to prepare students for life. The high schools did not want to offer their students only a college preparatory curriculum, unless that also was a strong foundation for life—and whether it was proved to be the subject of great debate. The Ann Arbor superintendent spoke for many high schools: "In developing character and fitting pupils for the actual duties of life," he wrote in 1874, "preparatory studies are probably not so fruitful of immediate results as high school studies proper."[61]

This conflict over the goals and purposes of the high schools was palpable and not easily resolved. The high schools, caught between the demands of the university and those of the local communities to prepare students for life, sought to balance these two expectations. These schools, aware of their connection to their communities and their reliance on local taxation for support, could not move radically beyond public needs and desires. Some high school administrators welcomed university inspections and recognition as a way to move recalcitrant boards and communities toward reform and change. Others understood that their communities had little interest in the university and reform. Angell was in no position to dictate to them, and they had no obligation to follow his lead. Michigan's schools and its university never completely resolved this tension, and it remained a significant part of the larger debate over the role of the schools in society and in relation to the university. The inspection program brought the two educational levels closer together and gave them a foundation for addressing this tension, but it remained a sensitive issue as the high schools struggled with the university's shifting entrance requirements.

# IV

Throughout the last decades of the nineteenth century, Angell's inspectors closely analyzed the courses that high schools offered, and they pointed out places where those courses did not meet university expectations. In school classrooms on a daily basis, the teaching may have been ineffective, but as long as a school offered classes that matched university expectations, that school likely earned accreditation. The point was to ensure that the schools understood their relationship to the university and organized their courses of study in response. A higher level of scholarship ideally would follow as the university increased its admission standards.

Whether the high schools were of high quality or not was secondary to the goal of creating and establishing the notion of high schools as preparatory schools that led up to the state university in Ann Arbor. The inspection program worked to establish an image and a tradition of secondary schools as being just that: secondary to the higher institution known as the university. Not every school was going to send students to the university every year or even most years, but Angell wanted these schools to begin to see themselves as feeder schools to the university. The annual inspections by professors helped to establish this tradition and to develop in the public's mind and in the lower schools the dominance of the university. After all, university professors entered the high schools, evaluated them, and passed judgment on them. The teachers and administrators in the high schools were not going to the university, knocking on its doors, and suggesting ways in which the university could do a better job of completing the high school course. Rather, the inspection program took note of the high schools and their work, and then used that knowledge as a basis for determining what the high schools still needed to do to prepare students for university courses. These inspections gave the university an edge in knowing what the high schools were doing and were capable of doing, and the university exploited this knowledge to raise its admission standards and to push the high schools to continually accept the elementary work that the university wanted to relinquish. The knowledge gained from these visits allowed Angell to ensure that the work of his university, no matter how rudimentary it may have been at the beginning, was always somewhat in advance of the high schools. The University of Michigan was establishing itself at the top of an educational system and making higher education—not secondary education—the terminus or final stage of education.

Angell's actions, however, underscored a new and fundamental role for the high schools in selecting the students who would enter the University of Michigan. While their standing in the educational hierarchy was challenged by the inspection program, the high schools gained the authority for certifying students for advanced study. Students could still enter the university after passing a traditional entrance examination, but more and more of them used the high school diploma to unlock the doors guarding the university's classrooms. The basis for admission was shifting from an examination where students exhibited their knowledge and mastery of subjects to a credential that signified that students had completed a prescribed course of study. While ostensibly certifying that students had learned something, the high school diploma primarily indicated that students had spent time in a classroom with teachers who were more or less capable and who had a credential of their own, preferably one from the University of Michigan.

In the years and decades to follow, other universities and colleges adopted Michigan's model and helped spread the credential system throughout the nation, but the high schools did not merely acquiesce in accepting this model and the growing dominance of the universities and colleges. The Bay City school may have let the dust settle over everything, but other schools were active and energetic in establishing their own priorities and programs, even though they never abandoned a role in preparing students for college. In Michigan and elsewhere, the high schools had to balance the dual functions of educating students for college and for the needs of life. The following chapters consider the rapid expansion of Michigan's model throughout the country and the reaction of the high schools to the growing authority of the nation's colleges and universities and to the dual demands placed on them.

# Chapter 3

# Michigan Launches a Movement for Regional Accreditation

## I

School administrators, university professors, and presidents used a number of metaphors to explain the lack of a clear educational hierarchy connecting elementary, secondary, and higher education. James McCosh at Princeton was partial to a staircase. Others hoped for a sturdy ladder to carry students from one level up to the other. The superintendent in Oshkosh, Wisconsin, wanted a "suitable and well graded road" to connect "the homes of the nation by way of the kindergarten, rural and elementary schools with the high school, and in turn the high school would connect with the professional school, the college and the university."[1] A number of university and college presidents throughout the country had been trying to grade that road. James B. Angell's lead in Michigan became a rallying call for these university presidents as they quickly stepped behind Michigan's innovative accreditation program. As they did so, they gave greater authority to the secondary schools for certifying and credentialing students for higher education. Angell had started the transition from an admissions system based on exams to a credential, and the spread of the inspection program solidified this approach throughout the Midwest and other regions of the country.

By the mid-1890s, accreditation programs had become the predominant method for admitting students to college, although there were notable holdouts to this trend among the nation's elite, private universities and colleges. As Michigan's model swept throughout the Midwest, university professors and high school administrators across the region—at times banding together to promote necessary reforms—worked to improve the secondary schools and to align them with higher education. As more universities embraced the program, they began to cooperate in accepting students from schools on the accredited lists of other universities. This cooperation led to a de facto standardization in entrance requirements, much to the benefit of secondary schools. These schools now found it easier to prepare students for a number of different colleges, since most institutions of higher education in a region came to accept students from any accredited school without demanding separate standards.

The South tried to follow the lead set in the Midwest, but, struggling with poor educational conditions following the Civil War, universities and schools in the region offered only a rudimentary education. Nonetheless, they implemented accreditation programs to articulate the two levels. However, unlike the Midwest, they did not develop a consistent approach to enrolling students without an examination. Some universities insisted on inspections before accrediting schools, while others recognized schools and accepted students without relying on a personal visit. New England's colleges also had begun accrediting schools without the important safeguard of an inspection that Michigan and other midwestern colleges demanded. Whether they required a faculty visit or not, these accreditation programs shared a key characteristic. They underscored the creation of a hierarchical system that put the colleges and universities at the top.

New England's approach to accrediting schools often found itself in conflict with the exam system favored by some of the region's elite colleges. It was in part because of this conflict over admitting students that New England formed the nation's first regional association of secondary schools and colleges in 1885. This association—and similar organizations that developed in other regions—provided necessary leadership in building an educational system at a time when no central agency existed to coordinate educational policy. With professors, presidents, and school headmasters gathering together, the New England Association created the foundation for eliminating differences both in admission requirements and in how colleges admitted students, and thus fashioned greater harmony between the two educational levels in the region.

Throughout the country, accreditation programs and regional associations started to build an educational system in the last decades of the nineteenth century. Some of the barriers between the two levels were slipping away. This chapter explores the emergence of regional approaches to developing such a system of education. It begins by focusing on the spread of the Michigan model throughout the Midwest and far West and the growing standardization it provided in the years between 1870 and the early 1890s. It then shifts to the South and that region's tentative steps in developing a similar system of education through different types of accreditation programs. Finally, the chapter ends by examining New England's response to Michigan's innovation and the development of the nation's first regional association of secondary schools and colleges.

# II

What united the many different universities and colleges throughout the country as they embraced the accreditation program was the shared need to join higher education with the secondary schools in a time of changing social, economic, and political priorities. Angell's innovative inspection and accreditation program in Michigan provided the model, one that some schools freely adapted to their particular needs. By 1896, according to the U.S. commissioner of education, 42 state universities and agricultural colleges and around 150 other colleges had implemented some sort of accreditation program.[2]

Not surprisingly, many state universities in the Midwest readily embraced Michigan's pioneering program for reasons that matched Angell's goals and ambitions. Faced with the same context as Michigan—the dominance of public high schools but their questionable ability to meet the standards set by university admission requirements—these universities accepted the promise held out to them by Michigan's example. A system of inspections and accreditation had the potential, many believed, to elevate the standards of the lower schools, send more students to higher education, and allow the region's universities to flourish as centers for advanced research and study. That goal certainly was the hope of the state universities in Wisconsin, Illinois, and Nebraska. The president of the University of Nebraska, as firm a supporter as Angell was, believed in the power of the inspection program to build an integrated state system of education. "Admission by certificate," he declared in 1893, "brings all parts of the school system together in a helpful and stimulating way."[3] By 1895, his faculty had inspected and approved seventy-five high schools. Most of these schools, however, were not capable of preparing students fully for the academic courses of the university. Nebraska's president was making progress but more needed to be done to bring his state's schools together "in a helpful and stimulating way."[4]

The University of Illinois also had made headway in inspecting and accrediting schools. By 1895, professors there had visited and approved 120 schools; a few years later, over 200 schools were listed on its accredited roster.[5] One professor who visited and inspected these Illinois schools underscored the significance of the relationship that developed between university professors and high school administrators as a result of this program. "School boards are often inclined to think that a school can get along another year," Stratton Brooks stated in 1901, "even though necessary improvements have been called to their attention by the superintendent." The university's support for ambitious superintendents, however, reportedly could induce these local school boards to embrace necessary reforms, such as adding classrooms or building new schoolhouses, furnishing libraries with numerous reference books, and supplying scientific laboratories with the best apparatus.[6]

This emerging relationship highlighted the pivotal role that professional educators—university professors and presidents and those secondary school administrators who trained in colleges and universities—were coming to play in determining the direction of the nation's secondary schools. One superintendent, for example, reported to his board that the university inspectors requested that the school purchase new scientific equipment. As a result, the board "immediately ordered the purchasing of $100 worth of apparatus and will soundly favor a like appropriation annually until we are fitted out for accredited work."[7] Another superintendent asked the university inspector to write a letter encouraging the board to make crucial changes in the high school. "I believe [such a letter] might be instrumental in doing much good," he said. "We have had a hard struggle here to secure four years in the high school, to obtain material with which to work, and to secure college graduates for teachers." This superintendent hoped that joining forces with the university would propel the local board toward establishing a strong, rigorous four-year high school program. "A few words from you might stiffen up the backbone of some [of] our board members along this line."[8] Throughout Illinois and the region, university professors used the inspection program to build coalitions with reform-minded principals and superintendents and to push for improvements in the high schools.

Other states modified this inspection program to establish their own approach to implementing reform and to improving the secondary schools. Minnesota's method of accrediting schools in the late nineteenth century was more complicated than Michigan's process. The high schools in Minneapolis and St. Paul, in part because of their size and quality, automatically earned the privilege of sending students to the university without examination. Other schools were at the mercy of a state high school board that consisted of the governor, state superintendent of public instruction, and president of the university. This board inspected high schools and provided the stronger schools with additional state aid and the privilege of sending students without examination to the university. Eventually, the

University of Minnesota agreed to inspect and evaluate all of the state's high schools and to allow the accredited ones to send students to the university without examination.[9] Minnesota's president lauded this approach to inspecting and accrediting high schools. "Our experience," he said in 1893, "has been decidedly in favor of the certificate [or accreditation] system. We find that our freshmen are better prepared for their work, and with the worry of entrance examinations and conditions removed they do better work in the university."[10] Indiana followed Minnesota's lead in developing a state board, which also included public citizens, to inspect and evaluate high schools.[11] Charles W. Eliot, Harvard's president, believed that such an "independent inspecting authority" was "greatly to be preferred" to the Michigan model, since it prevented universities from accrediting schools simply to bring in more students.[12]

The University of Wisconsin had little interest in Eliot's criticisms or Minnesota's state board. Instead, in 1876 it followed Michigan's lead and launched a program that was an almost exact replica of the model that Angell developed. Professors toured the state to evaluate schools and pass judgment on their quality, and the stronger high schools earned the right to send students to Madison without examination.[13] There was a significant difference, however. To ensure a steady supply of students, the University of Wisconsin and most other state universities in the region—with the exception of Michigan—opened their own preparatory departments.[14] These universities took students, prepared them in preparatory or sub-freshman programs, and then sent them up into the advanced courses of the university proper. John Bascom, the university's president, anticipated that the accreditation program would elevate the lower schools so that he could close his institution's preparatory program.[15]

These preparatory departments in Wisconsin and elsewhere significantly blurred the distinction between higher and secondary education, but they provided an essential route to college for many who otherwise would have had no chance of attending a university. In many states and regions the inadequate quantity and quality of high schools prevented students from having access to a strong preparatory program in foreign languages, mathematics, science, and literature. Especially for students from rural areas with limited access to well-equipped high schools, college preparatory departments offered one of the few avenues for getting the preparation needed to enter college. Outside of Madison and Milwaukee, for example, most Wisconsin students lived in rural areas, and Bascom worried that these students "would experience serious difficulties in reaching the University, if we refused them preparatory instruction."[16] About half of Wisconsin's students in the mid-1870s entered through this department rather than through the public high schools.[17]

While this department provided secondary students an opportunity to prepare for college, it offended Wisconsin's principals. They criticized the university for competing with the public high schools through this sub-freshman program. They saw secondary education as the proper work of their institutions, not of the university, and they urged the university's Board of Regents to abolish the preparatory program. "We feel that the graded schools of the state are justly entitled to protection at the hands of the Regents of the University, from being obliged to compete with its preparatory work," they declared in 1878.[18] As long as the university maintained what was, in effect, its own high school, the rest of the state's secondary schools suffered from the flow of students to Madison, these principals argued.[19] Bascom sympathized with the state's principals. He too wanted to end preparatory work and shift resources to advanced courses. Until he could do so, his university would not emerge as a research university, but Bascom maintained that few high schools in the state were yet able to prepare students adequately for any course in the university except the scientific course—a course that required fewer years of foreign language and classical studies and, thus, for many was less prestigious.[20] He did not "quite trust the assertion that the [preparatory] work will be done at once by the high schools if it is thrown upon them. We fear that there would be a fatal break in it, and one from which it might take years to recover."[21] The time had not come when Bascom could shift resources from the preparatory work to the academic departments, but he hoped the inspection program would help to bring that day about quickly.

Since the preparatory program had negatively affected interactions between the university and the high schools, Bascom needed to repair those relations. He used the inspection program, much as Angell had been doing in Michigan, to understand what the high schools were capable of accomplishing and what the public would support. He then relied on this knowledge in raising his entrance requirements. By slowly increasing standards in line with what the secondary schools could do and gradually encouraging them to do more, he reached a point in 1880 when he thought he safely could close the preparatory department, although he maintained a sub-freshman course in Greek for those students from schools without classical courses.[22] Charles Kendall Adams, who became president in 1892, understood the dynamic relationship between entrance standards and the accreditation program. It "tends to adapt the requirements of the university and colleges more perfectly to the possibilities of the high schools," he believed.[23] The inspection program gave Bascom and his successors an opportunity to work with the high schools and elevate standards, and, as the lower schools improved, to shift resources from the preparatory department to the advanced work of the university.

As reticent as Bascom had been in ending the university's preparatory work, he found that enrollment numbers remained relatively steady once the department closed and more schools gained university recognition.[24] By 1886, after the inspection and accreditation program had been in operation for ten years, the president concluded that the university "is constantly increasing its hold upon the high schools, and has now a large and influential list of accredited schools, which are shaping their courses of instruction in reference to it." Students in the university, he said, "are now all fitted, with the exception of a very small Greek class, in the schools of the state."[25] The university had made steady progress through the inspection program in using the public schools to replace the students who had been in the preparatory department, and, in the process, it molded the four-year high school as the primary route students took to get to college. In 1878, only three high schools (all in the state's larger towns) and one academy earned the right to send students to the university without examination, but the number of schools on the accredited list steadily increased over the years. In 1884, twenty-six schools appeared on the list (including one academy in Chicago). Three years later, the roster recorded sixty-two secondary schools.[26] By 1900, 80 percent of the state's one hundred and fifty high schools were on the university's accredited list, and most students entered the university without examination from approved high schools.[27] Some schools, in particular, worked hard to earn the university's recognition. The high school in Bloomington, Wisconsin, for example, had hoped for some time to gain a spot on the university's accredited roster. When Albert W. Tressler, the university's principal inspector, finally gave the school a place on the coveted list, the students in Bloomington erupted in joy. According to one account in the early 1900s, the students "paraded the streets carrying appropriate banners and scrimmaging to their entire satisfaction."[28] Bascom and Adams, through the accreditation program, had helped bring about this pronounced shift from the 1870s when most students entered the university through its preparatory department and when most high schools complained about university competition.[29]

To the south of Wisconsin, the University of Chicago, building from the ground up thanks to Rockefeller money and opening in 1892, took similar steps to create a vibrant system of articulated schools. William Rainey Harper, the university's first president, outlined a bold vision for education that included close relationships with private academies and public high schools. His program of "affiliated schools" placed a number of private schools under the educational control of the president and his Board of Trustees and essentially made them departments of the university. Harper and his faculty played a role in these schools by helping to appoint new teachers, who, along with their colleagues, then acted as examiners for the university, wrote examination papers, and administered the tests. Students who passed the exams entered the university without having to take additional examinations. Harper even hoped to create a junior college program where smaller colleges would affiliate with the University of Chicago and offer the first two years of the collegiate course. Students from these affiliated junior colleges would then matriculate into the higher classes of the university. Not surprisingly, most schools and small colleges were unwilling to give up such control and few acceded to Harper's ambitious but complicated plan. By 1898, only eleven secondary schools—including the Harvard School, the Princeton-Yale School, and the South Side Academy—and four colleges were affiliated with the university. At the time of Harper's death in 1906, the university had disbanded the affiliated schools program.[30]

Public high schools and those private schools that refused to affiliate had the opportunity to apply for "cooperating" status with the university. Under this scheme, the university essentially accredited specific teachers and made them deputy examiners for the university. These deputy examiners created their own entrance examination papers for their subjects, submitted them to the university for approval, and then administered them to their students. If students passed these exams, they could bypass the corresponding entrance tests administered by the university. This cumbersome approach to examining and accrediting teachers did not last long, and the university, reverting to the model prevalent throughout much of the Midwest, began to inspect schools and admit students without examination from accredited schools.[31] Fifty-four public high schools entered into this cooperative relationship with the university by 1898, including schools in Colorado, California, Ohio, and Kentucky, as well as in Illinois, Wisconsin, Michigan, and Minnesota. Fourteen of these schools were public high schools in Chicago.[32]

To encourage these schools to further develop their courses in line with the university's needs, Harper devised a number of additional programs and initiatives. He offered scholarships to students from the cooperating public schools as well as from the affiliating private institutions. He often invited those schools that affiliated or cooperated with the university to march in official academic processionals at the university in positions directly behind the university faculty. Harper also brought principals, other administrators, and teachers to campus to meet with university professors and officers. In these semi-annual meetings, "noted educators" gave lectures and discussed pressing educational issues, while teachers and professors met in departmental conferences to discuss issues specific to teaching English, science, history, and other subjects. As the director of the university's department of affiliations stressed, these conferences helped to bring the high schools and university into "closer personal touch and sympathy through a joint consideration of the problems of the work in which all are alike engaged."[33]

Finally, to solidify the university's influence on the secondary schools, Harper offered free or partial tuition to teachers from the affiliating and cooperating schools if they entered the university's new teacher's college. Although not solely for teachers, this college, which opened during the 1898–1899 academic year, catered to the city's school teachers by holding classes in central locations and after regular school hours. John Dewey, after he moved from the University of Michigan, and a number of other professors taught courses in philosophy, history, English literature, and classical languages. They also offered a few courses in pedagogy, but Harper wanted his new college to prepare teachers to earn the university's regular degrees. "The College for Teachers is not a normal school, but an arrangement of instruction intended to give those teachers who have not had a full college training the benefit of such training," he stated when announcing the new college.[34] For him, the real means of strengthening the area's teaching force and, thus, the students entering his university was not to offer training in teaching methods but to offer rigorous, demanding university courses in the academic disciplines. Armed with this knowledge, local teachers not only would be on their way to earning a bachelor's degree but also would have a strong foundation for teaching history, science, and other courses in their own schools. Such liberally educated teachers, Harper believed, would be filled with the spirit of a college education and thus encourage their students to pursue advanced education. By 1900, nearly five hundred students attended classes in the university's college for teachers.[35]

While not as far-reaching or ambitious in its plans, the University of California similarly championed a hierarchical educational system with it perched at the top. Although it did not implement its accreditation program until 1884, it developed by some accounts the most rigorous and efficient inspection program among the nation's universities.[36] California took the Michigan model and built in more safeguards to ensure that the inspections highlighted the strengths and weaknesses of the accredited schools. California dispatched professors in each of the core subjects— English, mathematics, history, classics, and science—to examine schools, while other universities settled for sending one or two professors to evaluate all of a school's offerings. "Our system of accrediting seems to us more thorough, more effective, more sure of good results than any other which has been adopted in our American colleges," the university's president declared in 1893. "We are not satisfied with a general impression gained by one or two professors in a brief visit to the school." California also required the high schools to send "specimen papers showing the style of [their] work," and it evaluated secondary schools based on how well their graduates did in college.[37] The state's high schools moved slowly in embracing this comprehensive accreditation process. Four years after launching the program, the university listed only six schools on its accredited roster. In 1892, the list contained thirty-one schools, but by 1901, it included over one hundred schools.[38]

By sending out specialists in each subject area, California ensured that its examinations of schools were searching and thorough. Wisconsin, Michigan, and most other universities did not send specialists for each subject area, and they did not require that the professors they did send confine their investigations solely to their particular specialties. As a result, the inspections often were somewhat limited in scope. Harvard's Eliot was highly critical of this approach.[39] As long as one or, at most, two specialists represented the entire university and inspected courses ranging from Latin to history to natural science, the program remained somewhat cursory, he argued. His analysis was correct. For example, Frederick Jackson Turner, one of the most famous historians at the turn of the twentieth century and a professor at the University of Wisconsin who often visited the state's schools, concentrated most heavily on a school's courses in history.[40] He had little expertise in other subjects, even though he was responsible for evaluating a school's entire course of study.

Still, this limited approach may have been of benefit to the schools. Rather than having to contend with a number of recommendations spread across all of their subject offerings, schools in Wisconsin and Michigan dealt with a distinct set of expectations in one or two subjects. Thus, they earned accreditation for the entire course of study without being subjected to the searching examinations that California required. This approach may not have provided opportunities for the schools to improve their courses of study based on the recommendations of specialists, but it did give them some freedom and flexibility in designing their courses of study independent of the university, while also receiving university recognition. Eliot was right: the inspections tended to be cursory, but their superficial nature benefited the secondary schools.

Schools from Michigan to California also profited from the spectacular growth of the inspection system. As more and more schools applied for inspection and accreditation, the process of sending out professors became onerous and demanding. As a result, many universities and colleges began to accept schools on the accredited lists of other universities without requiring their own, independent investigations. Minnesota, for example, admitted students, without examination, from accredited schools in other states. Thus, a student from Bay City, on Michigan's accredited list, could enter the University of Minnesota without examination, even though Minnesota's inspectors never visited Bay City.[41] This reciprocal arrangement was increasingly popular among universities and colleges, and it helped to ease the transition from high school to college for many students who wanted to attend public or private universities in other states and regions. Northwestern University in Illinois, for example, accepted students from schools on the accredited lists in Wisconsin, Michigan, Minnesota, Kansas, Iowa, Indiana, and California.[42] Michigan and Wisconsin also accepted students from schools on the accredited rosters of other state universities. As more schools adopted this arrangement, students applying to attend colleges in other states and regions no longer had to prepare differently to meet the diverse requirements of various universities and colleges, and secondary schools no longer had to expect regular visits from more than one university. By making it easier for students to enroll in colleges and universities across state lines without having to meet different requirements, this new arrangement gave schools greater flexibility, and it helped to relieve classroom teachers of the struggle to ensure that their students had read, for example, all of the different Latin and English texts required by various universities.

Similarly, in a state where the public universities had adopted the inspection and accreditation program, it was often the case that the other colleges and universities in that state accepted students from schools on the accredited list of the state university. Thus, Lawrence University in Wisconsin admitted most students without examination from schools on the approved list of the University of Wisconsin. Likewise, Olivet College and Albion College in Michigan accepted students from schools accredited by Angell's university.[43] The state universities, as Angell had hoped, were becoming the head of a state's educational structure, to the benefit of college-bound students throughout the region. Even though the spread of the accreditation system and the acceptance of another school's certificates did not prevent universities from publishing different admission requirements, this arrangement did reduce the pressure and stress on the lower schools to meet a number of diverse and confusing entrance standards. In essence, because schools no longer had to pay as much attention to the various requirements of a number of universities and colleges, a de facto uniformity began to exist in regions where a number of universities and colleges accepted students from schools on another university's accredited list. As long as potential collegians had attended a school that met the requirements of an accrediting university, they likely were able to enroll in any college in the region without having to pass an exam or meet institution-specific requirements. The road to college was becoming smoother and better graded.

# III

Southern universities understood how challenging it was to elevate the secondary schools and align them with higher education. These universities were in no position to adopt Harper's ambitious approach to articulating the schools or even to consider the system of education that Angell wanted for Michigan. Devastated by the Civil War, the South struggled throughout the late nineteenth century to build a system of secondary schools for white students, and, in the midst of Jim Crow laws and hostility toward African Americans, white Southerners often had no interest in supporting and usually opposed black education. A suitable tax base to adequately sustain public schools did not exist in much of the South, and where it did, few citizens had the desire to propose taxation for public schools, white or black.[44] Edward Joynes of South Carolina College put it succinctly. "Our Southern States are almost wholly without secondary schools."[45] Colleges and universities in the region, consequently, opened preparatory departments, and few of these "colleges" enrolled students in advanced courses. The dividing line between higher and secondary education was the most indefinite of any region in the country.[46] Defining the boundaries between the two educational levels was a pressing concern but it was less important than establishing public support for education, building more public schools, and enrolling students in higher education. Even in the Midwest and West, the need for students pressed on the universities, but this demand was particularly challenging in the South.

The educational situation improved in the last decades of the nineteenth century, and the South's white universities and colleges started to adopt accreditation programs in the 1890s as a way to work with the lower schools and enroll students. Not all southern universities and colleges, however, embraced the accreditation program as Angell and his counterparts in the Midwest had developed it. Although some institutions in the South relied on inspections as a means of accrediting schools, others simply enrolled students when they came with a recommendation from their principals certifying their ability to continue their studies. This "certificate system," as opposed to the inspection program, did not send professors into the secondary schools before accrediting them. In their quest for students, not all colleges in the South were rigorous in their demands or expectations.

Tulane University in New Orleans in the early 1890s began to admit students on certificate and without examination from any school that adopted the university's recommended courses of study, employed competent teachers, and examined students thoroughly. Tulane granted such a privilege to a school only after one or more students from that school had successfully passed the traditional entrance examination. Finding schools that aligned with university courses was not easy, and by 1895, Tulane listed only six schools on its accredited roster.[47] By 1902, the list had expanded to only fourteen affiliated schools.[48] The University of Alabama had even fewer standards for approving schools and by the early 1900s had accredited twenty-eight schools to send students to the university without examination.[49] Johns Hopkins University in Baltimore imposed some higher requirements and accepted students on certificate only in certain subjects. Students wishing to enter this university still had to pass exams in a number of subjects, including trigonometry and Latin and Greek prose composition. They also had to sit for exams in some aspects of French, German, English, and science. Certificates to Johns Hopkins did not free potential students from much of the burden of preparing for and taking entrance examinations.[50]

The University of Mississippi also admitted students without any further examination, but it did so only after university professors had visited and inspected the secondary schools. In 1895, just a few years after launching this program, the university accredited fifteen schools; a decade later it listed sixty-eight affiliated schools.[51] Before turning to this program as a way to admit students, however, the university in 1874 had opened its own high school to teach preparatory courses. Mississippi's actions in operating such a high school and focusing on preparatory work were not unusual. Throughout the South, colleges and universities in their need for students developed preparatory departments or enrolled students directly in their institutions without strong secondary preparation. With most of their students in preparatory classes, these institutions were colleges in name only. The teachers in the South's academies and public high schools maintained, as many had in Wisconsin, that these actions interfered with their work and hindered their full development as strong secondary schools. "All that we ask," according to R. Bingham, principal of the Bingham School in North Carolina, "is a fair field and no favor. Raise the standard for admission into the colleges as at the North; or for exit as at the University of Virginia; or for entrance and exit; exclude children and mere boys by limit of age—say sixteen or seventeen; and there will be preparatory schools enough to do all the work without any endowment but brains."[52]

The colleges slowly came to understand that their preparatory work negatively influenced educational development in the region. As one professor at Washington and Lee University put it, "When the colleges take these boys who ought to be in school a year or two longer, they are killing the educational goose that lays the golden eggs from which college students are hatched." He counseled colleges to simply shut down their preparatory programs, even if that action temporarily reduced their enrollment numbers. "Experience has shown that the way to cut off preparatory classes is—to cut them off," he insisted in 1900, "and have the nerve to withstand the outcries of alumni and friends who exclaim against such a policy because it reduces numbers."[53] Only in this way, he implied, would the South be able to strengthen its lower schools and, as a result, its universities.

Mississippi, in particular, recognized that it would have to abandon its preparatory work completely if it wanted the state's high schools to improve and provide a foundation for advanced university work. By 1883, according to the former chancellor, "the board of trustees, deferring to that objection and considering the further fact (more weighty by far) that the academies of the State had begun to recover from the ruinous effects of the late war and to do much more thorough educational work than had been done before, abolished the school."[54] Although the university closed its high school, it apparently kept open a preparatory department that provided some secondary courses for students who could not meet the requirements for admission into the university. By 1892, it abandoned even this work. Its inspection and accreditation program, launched that same year, was central to its efforts to focus on collegiate, as opposed, to secondary work. Paul Saunders, a professor at the University of Mississippi, underscored the value of this accreditation program. He revealed that many of the state's superintendents readily were meeting the university's requirements so that the affiliated list in 1897 included forty schools that "were giving courses in at least three topics of sufficient advancement to prepare for the freshman class of the university."[55]

The University of Tennessee similarly inspected schools and admitted students without examination. However, it did not absolutely require that schools submit to an inspection. Professors there had the option of evaluating a school through correspondence with local officials. Whether requiring an inspection or not, the university's accreditation program—which had approved thirty-five schools by 1902—made a difference in the educational system of the state, the university's dean believed. Admitting students through such a program, he claimed, was "a better test of the candidate's qualifications." Many supporters of the program in other states concurred with the dean that "a worthy teacher who has had [a student] in hand two or three years is a safer guide in this matter than the results of an examination. He knows not only what the boy has done, but his capacity, tastes, and habits, and therefore what he probably will do."[56] The results of the program did not undermine the dean's conclusions. The accreditation program, he wrote, "has given us as a rule better prepared students." Not only did it provide stronger scholars, he argued, but it also brought the high schools more into line with the university.[57] An official from the

University School in Nashville agreed with the university's dean. "The certificate system is a powerful tonic to schools," he claimed. "Those that have the privilege of entering their graduates without examination strive to retain it. Those that have not this privilege strive to win it. The lack of the privilege is a badge of inferiority that schools do not like to wear."[58]

Hoping for the same results, the University of Arkansas required school inspections before granting the privilege of sending students without examination. Unlike the trend among many universities in the North, however, this university, along with other southern universities, often admitted students who had not completed the full high school course of study and graduated. The University of Arkansas excused these students from taking entrance examinations in all of the subjects they had studied in an accredited high school if they had a certificate from their principals.[59] The University of Tennessee dissented from this approach. Officials there hoped that the inspection program would keep students in school longer by removing the obstacle of an entrance examination. "The desire to win a certificate" and not have to study for the entrance examination, argued an educator in Nashville's University School, "will sometimes cause a boy to remain at school until he is well prepared, when he would otherwise try to enter college, although burdened with conditions."[60] The universities often complained that high school students lacked the necessary preparation when they entered college. For some universities, the accreditation program sought to ensure that students remained in high school for four years and gained a thorough grounding in preparatory subjects. Thus, graduation from a secondary school became a key prerequisite for entering some southern white colleges.

The educational situation was even more precarious for black students and their schools. Since white Southerners refused to integrate black students into white schools, the region maintained a segregated system of education. This dual system—growing out of white racism and hostility toward the former slaves—only heightened racial divisions in the former Confederacy. What efforts whites expended on education—and for many years, these efforts were minimal—focused on white students. "There is no nobler race than the real Anglo-Saxon of the South, no race capable of higher cultivation and greater achievements," maintained R. W. Jones, a University of Mississippi professor. Expending valuable resources on black education hindered efforts to elevate the white race and angered Jones and other white Southerners. "It is true that we labor under many disadvantages," he wrote in 1900. "We have to educate two races side by side in separate schools; the white race of the South has to carry well nigh the whole of this double 'burden.'"[61] He was wrong—southern whites usually siphoned money from African American schools to strengthen white schools—but his complaints constituted good propaganda, established an excuse for the poor state of white schools, and fanned the flames of hostility toward African Americans.[62] Only grudgingly, then, did southern whites even allow African Americans to develop schools and colleges.

With white southern hostility and few opportunities for formal education available before the Civil War, the black community in the South struggled to construct a strong common school and college system in the late nineteenth century. By 1900 only 36 percent of African American children between five and fourteen years of age attended school. The situation was even worse in the secondary schools. In 1890, only 0.39 percent of African Americans of high school age enrolled in secondary school, rising to only 2.8 percent by 1910.[63] Northern philanthropic agencies—including the American Missionary Association—and the Freedmen's Bureau during Reconstruction offered some support in building schools and colleges in an effort to get more black children into school, as did considerable efforts at self-help. But, in the absence of a strong tradition of common schools, the region's black colleges—like many of the South's white colleges—had to focus primarily on secondary and elementary instruction. In the late nineteenth and early twentieth centuries, most black students who attended a college or university were in preparatory courses. This situation was true of some of their northern white counterparts, of course, but, like most white universities in the South, black institutions for higher education often were colleges in name only.[64]

Leland University in New Orleans, for example, initially had to provide elementary and secondary education when it opened its doors in 1870. Founded with northern philanthropic support, this university, although its charter did not discriminate by race, primarily enrolled black students. As the college grew in stature and passed from being a preparatory school to a college with advanced courses, it developed an accreditation and affiliation scheme in the early 1890s that resembled Harper's plan in Chicago. Leland professors crafted the course of study for the university's affiliated high schools and were instrumental in appointing teachers to the lower schools. The university even paid the salaries of the teachers in the secondary schools, thus binding the university and the affiliated schools in both educational philosophy and financial matters. Graduates from the affiliated schools automatically entered Leland University.[65] Howard University in Washington, D.C. similarly reached out to the secondary schools, although it did not embrace such an ambitious scheme. It began to work with the city's secondary schools through its own accreditation program, and by 1895 admitted graduates of the city's high schools without examination.[66]

Two decades after the North first developed these programs, some white and black colleges and universities in the South slowly began to implement their own accreditation programs. Not all adhered to an inspection component, and some colleges—desperate for students—simply admitted pupils from any school based on a principal's certificate. Educational standards in the South remained low in comparison to colleges and secondary schools in other regions. Accreditation programs, however, were beginning to provide a way to improve these schools, to define the work of each educational level, and to align higher and secondary education in a system of schools. Importantly, southern colleges helped to establish the four-year high school course as the standard requirement for admission to college. Not until after 1910, however, did white colleges and universities in the South band together to develop a widespread system of accreditation that matched the spread of the program in the Midwest and New England. It was not until the 1930s that black colleges and secondary schools implemented a parallel program to inspect and accredit schools.[67]

# IV

The South struggled to build a system of education, but New England's prestigious colleges and universities, conversely, had the advantage of relying on Exeter, Phillips Andover, Boston Latin, and a handful of other superior schools and academies. Harvard, founded in 1636, had well-established traditions and relationships with preparatory schools. In contrast, public universities such as the University of Wisconsin, founded in 1848, and most universities in the South lacked such hallowed histories. Universities in these regions were growing and creating systems of education in very different contexts. The smaller colleges in New England, however, understood the challenges that Wisconsin and Mississippi faced. They, too, needed students, and lacking the relationships with preparatory schools that their stronger counterparts possessed, had to find ways to attract students to their campuses. While Harvard and Yale adamantly refused to relinquish their entrance examinations and, thus, control over who entered their ivied halls, Brown, Wesleyan, and Amherst, along with many of the region's other colleges, embraced a different model for admitting students and for strengthening the connection between the secondary and higher levels. They followed Michigan's lead in accrediting schools and admitting students on the basis of a secondary school credential, but they did so with a crucial variation.

Those New England colleges that supported an accreditation program instead of the examination system almost always did so without insisting on faculty inspections. As was generally true in the South, these colleges usually admitted students solely on the certificate or recommendation of a high school principal. There were few safeguards to ensure that the accredited secondary schools were of a high quality and were capable of sending well-educated students on for higher study. The colleges might examine a school's course of study to ensure that it aligned with their own courses, but few colleges undertook even this rudimentary analysis—a review that took place in college offices rather than in high school classrooms. Some New England colleges did impose one safeguard on the accreditation process. They evaluated secondary schools based on the performance of students from those schools in the first year of college. Schools that regularly sent well-prepared students who succeeded in college remained on the list of secondary schools that could send students without examination. Brown's president claimed that his institution received certificates "from such schools only as our own experience, or other trustworthy information assures us are accustomed to do thorough work."[68] The president of Williams College, however, made it clear that expelling a student or dropping a school was generally unwise. Only in extreme cases, he pointed out, would the right to send students on certificate "be withdrawn as it has been in the past."[69] New England's colleges often claimed that they accredited secondary schools based on the performance of their students in the first semester in college, but these colleges rarely denied a school accreditation based on weak student performance.[70]

This lack of strong safeguards and standards in his own backyard only increased Eliot's opposition to accreditation programs. In 1890, he called the certificate system the "feeblest way" of admitting students to colleges and universities. He disliked Angell's model, but had even less respect for the approaches taken by his colleagues in New England.[71] Nonetheless, even Eliot eventually created a program at Harvard modeled loosely on the Michigan innovation. Harvard had a network of strong preparatory schools, but Eliot understood the importance of developing stronger relationships with the public high schools, especially as they became the dominant form of secondary education in the 1880s and 1890s. He remained determined, however, to build his university on the basis of students admitted through individual examinations. Eliot's solution was to embrace an inspection program, make it more rigorous than the Michigan model, and withhold any privileges or benefits to students from accredited schools. Starting in 1891, his program sent six professors, each trained in the core academic subjects, to inspect schools, analyze them carefully over a number of days, and provide recommendations for improvement. These schools benefited from a closer relationship with the university and from the expertise of Harvard faculty, but they never earned the privilege of sending their students to the university without examination. Eliot placed his faith in examining students to see if they had actually learned anything in secondary school, but he was willing to send his professors into the schools to build stronger relationships with them and improve them.[72]

Eliot's opposition to admitting students without examination, even as he built stronger relationships with the high schools, underscored the tension in New England between the traditional method of admitting students by examination and the newer certificate system. The principal of the Hillhouse High School in New Haven, Connecticut, maintained that the certificate system was preferable to the traditional examination, in part, because it challenged the secondary schools to do exceptional work. "Is it not clearly evident," he asked in 1892, "that admission by certificate makes us personally responsible for good work?" The obligation for certifying a pupil for advanced study put the principal's reputation on the line. The secondary schools, rather than the colleges, determined which students would have the opportunity to continue their education in college. Thus, the principal was more apt to take this responsibility seriously, Hillhouse's administrator concluded, and ensure that students had prepared adequately for college. "If my honor is involved in my recommendation, then, indeed, I must be careful." Based on observing the two systems operating simultaneously for seven years, he concluded that better work came from the students who needed a principal's certificate to enter college than from those who studied for exams. Such students worked hard to earn the principal's respect.[73] For the Hillhouse school, the certificate system proved the better admission choice, primarily because it demanded superior work in line with college expectations from the principal, teachers, and students.

John Tetlow, headmaster of the Girls' Latin School in Boston, argued that the examination system, rather than the certificate system, increased the standards of the secondary schools. He did not oppose the certificate system. After all, he sent three-quarters of his students to college on the strength of his recommendations and certificates. However, he recognized that the presence of examinations at Harvard and Yale helped to elevate the standards of the secondary schools and made the certificate system work. "So long as Harvard and Yale,—the two oldest and strongest New England colleges—refuse to admit students by certificate, and insist on a rigid entrance examination as a test of qualification for their courses of study," he maintained, "so long will the colleges which admit students by certificate continue to receive the benefit of the examination system." Colleges admitting on certificate, he implied, profited from the high standards that Harvard and Yale demanded through entrance examinations. Since most of the region's preparatory schools sent some students to Harvard or Yale, they developed their courses of study to meet the demands set by two of the most prestigious universities in New England. These standards and expectations applied to all students in the schools, Tetlow asserted, and not just for those going to Yale or Harvard. As a result, Brown could admit students on certificates and generally know that the students were going to be well prepared for university work. He predicted that "if the entrance examinations were wholly abolished, and admission by certificate were universally substituted in their place, the certificate system would be speedily and completely discredited."[74]

Tetlow had a strong supporter in Eliot, who had no intention of embracing the certificate system and shifting the responsibility for admitting students to the secondary schools.[75] Entrance examinations were particularly effective, he believed, in forcing the lower schools to adapt to the university, and Eliot prized the role exams played in creating a system of preparatory schools tightly aligned with his institution. Throughout the late nineteenth century, Harvard's needs shifted continually, in part to reflect its changing educational philosophy, and the university adjusted its entrance requirements and admission examinations accordingly. Schools that prepared students for Harvard, especially some of the private academies that acted as preparatory schools, quickly adjusted to the new requirements and examinations.[76]

William Collar, headmaster of the Roxbury Latin School, recognized the power of Harvard's exams to reform educational practices. "I have seen," he declared in 1891, "a profound change wrought in schools that prepare for Harvard College, in the teaching of Greek, solely from a different mode of examining." He went on to note that changes in the examination in physics also had brought about significant shifts in how preparatory teachers taught this subject, and this "revolutionizing" of teaching, as he put it, was not confined solely to Greek and physics. "A change not less remarkable nor of less educational value, in the teaching of geometry," he concluded, "can be traced directly to the influence, not of altered requirements, but of a decided change in the character of the papers set at the Harvard examination."[77] His experiences led him to conclude that the examinations had transformed teaching in the secondary schools. "The educative value of college examinations is very great," he allowed.[78] The headmaster of the Cambridge Latin School understood Collar's position. "I do not see how the colleges can in any other way tell so well what they want, as by the papers they set," argued William F. Bradbury.[79] Unlike the public high schools in Michigan and Wisconsin, these secondary schools had long traditions of preparing students for higher education, and they educated their students in academic courses that aligned with Harvard, Yale, and other colleges. To maintain their standing in relation to higher education, they had to adjust to the new requirements and standards.

Examinations, not certificates, provided them with the best way to understand what the colleges expected.

Some educators, however, believed that the examination system worked against the interests of higher education, since it posed an unnecessary hurdle that students had to cross on their way to advanced study. Horace Willard of the Vermont Academy claimed that the prospect of preparing for and passing an entrance examination discouraged some of the "most mature and thoughtful minds" from attending college. Examinations, he went on, led to cramming rather than to thoughtful preparation and rewarded those who "are characterized by acuteness rather than depth." Not surprisingly, then, he favored certificates as a more effective means of encouraging students to prepare for and enroll in college. But he also thought that the certificate system would show secondary schools that the colleges and universities trusted them and that this confidence, in turn, would lead to better work in the secondary schools.[80] The president of Wellesley concurred with Willard, and she argued that the certificate system "would strengthen the hands of the teacher, raise the character of scholarship, increase the interest in a college course, and promote the co-operation of schools with colleges."[81]

Dartmouth agreed and saw value in the certificate system, which it instituted in 1876, both for encouraging comprehensive work among students and for aligning preparatory study with college requirements. In order to prepare students for college examinations, preparatory schools often had to devote much of the last year to a review of previous material. This review and cramming in the last year meant that students were not spending time learning new material. To offset this problem, some colleges instituted preliminary examinations that students took prior to their last year in school. As a result, students then spent part of their final year preparing in new subjects rather than reviewing old material. Dartmouth claimed that this preliminary examination was only partially successful in alleviating the problem, since students still had to review old material not covered in the preliminary test. The certificate system, on the other hand, allowed secondary school students to spend their last year concentrating fully on new subjects; it eliminated the need to prepare and review for an entrance examination. "A teacher who looks forward to sending his pupils on certificate, can arrange his course without the drag of final, complete reviews," a Dartmouth professor stated in 1879. "There is no retracing the steps, no swinging round a circle; work done is done; both scope and method are broadened." Although he had seen this method of admitting students in practice for only three years, he nonetheless felt that secondary school teachers were becoming more zealous and were aligning their work with college expectations. Since Dartmouth evaluated a school based on how well students from that school did in college, it essentially judged the teachers in the secondary schools. This judgment encouraged the teachers to do better work, he believed. As a result, he concluded, there had been "a noticeable improvement in the preparation of those thus entering" Dartmouth through the certificate program.[82]

When it came to admitting female students specifically, some educators contended that certificates were the only option. For one president, admitting women by certificate was preferable to expecting them to sit for an examination. The president of Smith College declared in 1892 that it was not "wise on physical grounds to subject women to examinations at fixed times. It involves a useless nervous waste. It is far better to accept the testimony of teachers."[83] This attitude was popular in the late nineteenth century. Edward Clarke, a Harvard medical professor, had famously contended in the early 1870s that the mental exertion required by academic study would hinder women's abilities to reproduce and fulfill their function in society as warm, caring, nurturing figures. He and other educators doubted that women's health and their crucial role in the country would withstand the rigors of academic training.[84] Nearly thirty years later, James Taylor of the Chauncey-Hall School agreed that the "women's colleges of New England are very wise in accepting young ladies on certification." Female students, he claimed, "excel in marks, in fidelity, in time given, although going on certificate, the work of the boys who are expecting examination—until about the first day of June. Then the girl take[s] life easily, as she ought to, as it is desirable that she should. She is not worried with the strain and uncertainties of the forty-eight hours' test."[85] Examinations only pressured female students and taxed their weaker nervous systems, these educators claimed. Certificates eliminated this unhealthy strain and, therefore, were better for female students. Francis Waterhouse, the head of English High School in Boston—the nation's first public high school—challenged this viewpoint. "As to the waste of nervous energy of girls by examination, I regard much feeling of this kind as a false sentiment," he asserted in 1892. "Girls can be trained to take examinations without undue strain."[86]

Whether for women or for men, most colleges in New England embraced some sort of accreditation program and fell in line with the general trend prevalent in higher education in other regions. These colleges, however, rarely concurred on the standards for accrediting secondary schools or on specific admission requirements. Moreover, they did not agree on the type and amount of information they needed before accrediting a school and admitting students. Some asked for detailed accounts of the precise content of each subject studied and the textbooks used; other colleges only wanted to know whether a student had passed certain courses.[87] Secondary school teachers and administrators, therefore, spent much of their time filling out a variety of forms for all of their students applying to one or more colleges. Compounding the problem, not all colleges recognized such documents from every secondary school in the region.[88] Brown might accept certificates from a school in Rhode Island, but Amherst might refuse to accept students from schools in that state. If Rhode Islanders wished to attend Amherst, they had to prepare for that school's entrance examination. Admission by certificates was popular, but the certificate system lacked a regional coherence or unity of its own. As such, it failed to provide a basis for standardization as the inspection programs were doing in other regions.

Further complicating the situation, Harvard and Yale resisted the trend toward certificates and maintained their own entrance examinations. Although Columbia and Princeton then were less prestigious than Harvard and Yale, they too relied on entrance examinations and drew students from New England.[89] Inspection and certificate programs spread to many of the nation's universities and colleges, but entrance exams remained a significant challenge for those secondary schools in New England that enrolled students hoping to attend some of the most respected universities in the region and the surrounding states. Moreover, those colleges that admitted by examination had yet to settle on any standard or uniformity in admission requirements and exams. Cecil Bancroft, the principal of Phillips Academy in Andover, Massachusetts, protested this diversity in entrance requirements and examinations. "There is at present," he declared in 1885, "a very considerable diversity of nominal requirement in our good and reputable colleges, and I think it is not invidious to say that there is still a greater diversity in the actual requirement." This lack of uniformity posed problems for students preparing to take exams. "Two colleges setting examinations in precisely the same subjects," he protested, "do not set papers of equal difficulty and scope, nor mark according to the same scale."[90] Schools with students wishing to enter more than one university or college had to prepare their students for different sets of requirements and examinations. Waterhouse of the English High School in Boston suggested that all secondary schools might prepare students to pass the Harvard examinations. If they had the education to pass this exam, he concluded from his years of experience, students would be capable of passing any entrance exam. He doubted, however, that the colleges would agree to such a standard. They are slow, he claimed, "to adopt directly measures which indirectly they approve."[91] This reality only heightened the challenges facing the secondary schools in preparing students for college.

Across New England, the requirements for admission and the methods of enrolling students differed from college to college in the years before the twentieth century. Neither the certificate system nor the examination system had succeeded in reducing the pressures on New England's secondary schools. The diffuse nature of the certificate system, compounded by a diverse set of requirements for entrance examinations, meant that the region's secondary schools struggled to prepare students for college under stressful and taxing circumstances. Moreover, both approaches failed to encourage the colleges to seek a better understanding of the pressures and demands they placed on the secondary schools. In Michigan, Wisconsin, and other states in the central and western parts of the nation, the accreditation program was a dynamic, two-way process. In New England, the secondary schools, conversely, had few opportunities for interaction with higher education through the region's accreditation programs. The secondary schools filled out certificates and gained an understanding of what some of the region's colleges expected from them, but the colleges often had little sense of what the secondary schools were doing. The examination system was no better. Colleges set the standards and expected the lower schools to meet them. Higher education expected a lot from the region's secondary schools, but the colleges and universities rarely reached out to the lower schools.

Ironically, in a region that had the closest alignment between colleges and preparatory schools, the tensions between higher and secondary education were significant. By tradition, many secondary schools fed into higher education and had little choice but to respond to the shifting needs of the region's colleges and universities. This tradition, however, did little to reduce their frustration over the diversity in certificates, examinations, and admission requirements.

# V

New England's secondary schools—frustrated in part by the lack of consistency in methods of enrollment—proposed the formation of an association to bring about some uniformity in admission requirements and to effect a standard for entrance examinations and certificates. Early in 1884 and again in 1885, representatives from the Massachusetts Classical and High School Teachers' Association called for a meeting with their counterparts in the colleges and universities. They hoped that such a meeting would alleviate the serious "evils incident to the want of understanding and effective co-operation between the teachers of preparatory schools and the faculties of colleges."[92] The colleges and universities did not agree to meet with the lower schools until October 1885, but what emerged from this initial meeting was the New England Association of Colleges and Preparatory Schools—an organization that reached beyond Massachusetts to become the nation's first regional association made up of representatives from higher and secondary education and dedicated to reducing the gap between the two educational levels. In the absence of centralized control, either at the federal or regional level, this association provided an opportunity for higher and secondary education to begin to discuss challenges and craft initiatives to strengthen the region's evolving system of education.

Throughout its early history, this association devoted much of its time to resolving questions central to enrolling students by entrance examinations and certificates. The principal of Phillips Academy identified the problem during the association's first meeting. Regardless of whether colleges admitted students by certificates or examinations, he pointed out, they had been unable to agree on what should constitute a proper education. As American society underwent a marked transformation in the late nineteenth century, the needs of society and the role of education in preparing students to meet those needs continued in a state of flux. College representatives did agree that higher education, as well as the secondary schools, for that matter, should concentrate on a liberal education—as opposed to a vocational or more practical education—but what that meant differed depending on who was speaking and on what college was recruiting students. "There is no doubt that the chief reason for a diversity of requirement is the different estimate of what constitutes preparatory education, and of what constitutes a liberal education," Cecil Bancroft, the academy's principal, declared in 1885. The colleges had their own idea for what they should teach and expect from the lower schools, and this objective often depended on the particular circumstances of a specific institution. "The requirement for admission to a given college," Bancroft revealed, "is partly the product of the forces working at large in the educated and educating world, and partly the result of local necessities and limitations, together with personal theories and experience."[93] These local necessities and personal theories resulted in the problem that secondary schools faced. The requirements for admission to college and the exams and certificates that guarded the doors of higher education rarely meant the same thing for two or more colleges.

Whether he intended to or not, Bancroft highlighted a crucial challenge to greater uniformity in admission requirements. To bring about this standardization required colleges to fix a curriculum that represented national or regional ideas and norms rather than the peculiar circumstances of a local society. Even more rooted to their immediate communities, the secondary schools had to agree on some sense of uniform standards that applied throughout the country and that also addressed local expectations.

At the heart of the challenge between the two levels, then, lurked a debate between national or regional needs and local desires. Americans were moving to both coasts and settling the vast country between the two oceans. Tight-knit communities that had governed and organized much of life in the early American republic were being consumed by larger cities, factories, and corporations. Schools were not immune to this transition. Local needs could and did still account for some of what occurred in schools, but the growing expectations for uniformity and standards in schools, as students traveled from region to region, dictated that the schools pay attention to regional and national needs. The New England Association crossed state borders, in part because the region was closely connected geographically but also in recognition that the country was becoming interdependent in ways that bridged artificial boundaries. Balancing local traditions with the demands for regional and national uniformity—both to meet the challenges facing society and to relieve the pressure of preparing students for various universities and colleges—was not an easy task. Higher education struggled in the same context to agree on what a student should know to be able to succeed in college and in life. Constrained by their own traditions and aims, New England's colleges found it difficult to agree. As Bancroft said, "Perhaps we ought to wonder that the points of agreement are so many, the variations so few."[94]

Many educators hoped that this new association of New England colleges and secondary schools would reduce the tension between national and local needs, find a way to standardize requirements, and provide a measure of relief for the secondary schools. As an initial step, even before finalizing a constitution, the association's delegates asked the colleges to develop greater uniformity in admission standards by establishing consistent definitions for entrance requirements in key subjects. They further hoped that the colleges would accept the results of each other's entrance examinations and consider developing a joint examination board, "whose duty it shall be to set papers, conduct examinations and issue certificates on their behalf, which certificates shall be good in any college in the syndicate."[95] Thirteen colleges in New England—including, notably, Harvard and Yale—responded by uniting as a Commission of Colleges in New England on Admission Examinations to consider the wishes of the lower schools. In practice, the relationship between the two organizations was fairly cumbersome and deliberative. Quick action was rare. The New England Association passed a series of resolutions in its annual meetings, which a "committee to confer" then conveyed to the Commission of Colleges. This commission generally established a committee to consider the association's recommendations and to report back at the next meeting, usually the following year.[96]

This process was slow and at times frustrating for the secondary schools, but the commission managed to meet some of the hopes and desires of the lower schools. Their first interaction, however, did not represent a stunning success for the secondary schools. After meeting to consider modifications in entrance requirements in English, the New England Association requested the Commission of Colleges to unite in a conference with secondary schools to consider altering requirements in this subject. The commission agreed but invited only "a few preparatory teachers from different parts of New England." Of the fifty-six invitations sent, only twenty-three went to preparatory teachers. These secondary school representatives hoped that the colleges would reduce the number of authors required, accept more American literature, and resist making annual changes in the required books. Continually changing authors and books made it difficult for schools, especially smaller ones in rural areas, to adopt new books annually and find the resources to pay for them. The secondary schools further requested that the colleges not ask students to take a composition exam but instead accept forms certifying that the students had read the required books. The secondary schools then would not have to spend time in the final year reviewing books read in earlier years. They could concentrate, instead, on new literature and better prepare students for advanced study. The English professors, however, rejected most of the recommendations from the New England Association. They were not "ready to recommend radical changes in the general scheme of requirements in English and in the method of examination," the secretary of the commission reported, but they did agree to add more American authors and to keep a portion of the required reading list similar for at least three years.[97]

Within a few years, both the commission and the New England Association took up this issue again, in connection with the Association of Colleges and Preparatory Schools in the Middle States and Maryland— another regional association that began in 1887. The English professors seemed more inclined this time to meet the needs of the preparatory schools. For one, they agreed, in place of a formal entrance examination, to accept notebooks from students that included essays and compositions on the required books. They also asked the schools to prepare students in English grammar and to require students to read heavily in English poetry. Finally, they formed a committee of both higher and secondary education to devise lists of required authors and books for future years—some of which would not change annually. The secondary schools had gotten some of what they asked for initially from the English professors, and the addition of another regional association meant that the recommendations began to secure greater uniformity in English requirements beyond New England.[98] Tetlow was convinced that "to the organization of [the

Commission on Colleges] we owe the uniform requirements in English for admission to college which now generally prevail."[99]

The commission also moved to effect some standardization in the admission requirements for Latin and Greek. A committee designated by the commission to study the classical requirements conferred with "some of the prominent headmasters of New England preparatory schools" and detailed specific entrance requirements in both subjects, although the committee carefully pointed out that these were not prescribed courses but only suggestions.[100] The commission did encourage the colleges that required an examination in the translation of Homer and Vergil to agree on a uniform test. Eliot previously had made such a proposal, but neither he nor the commission was ready to argue that the colleges and universities should establish uniform requirements. No one was willing to infringe on the right of each college to set its own standards. Where they did agree, however, the colleges might use a common exam, the commission suggested. Furthermore, where colleges admitted on certificate, the commission recommended, they should agree on the authors and books required.[101] These recommendations represented important strides in unifying requirements and strengthening the connection between the higher and lower schools, but the commission overall remained cautious in meeting the needs of the secondary schools. The commission was hesitant to impose any requirements on the colleges, which were, after all, independent entities.

In a significant defeat for the secondary schools, the commission refused to sanction the idea that the colleges should accept any student who had completed a strong secondary school course of study, regardless of whether the subjects taken aligned directly with a college's specific requirements. The New England Association advocated such a change in the hope that it might bring the nonclassical high schools and the colleges closer together. Without strong classical courses, the public high schools and some academies were at a disadvantage, in comparison with the preparatory schools, in preparing students for college. A broadening of the subjects accepted for college would make it easier for students from the public schools to enroll in New England's colleges. The commission was willing to propose some alterations in college requirements to meet the needs of the high schools, but it remained timid in proposing such revolutionary changes. The New England Association, the commission declared, "proposes such a radical change in the college course that the Commission does not regard it as expedient to make any recommendation to the colleges on the subject."[102]

Gradually, however, New England's colleges embraced a degree of uniformity and standardization in some admission subjects previously unknown in the region. The secondary schools could breathe a sigh of relief, but the gap between higher and secondary education had not completely disappeared. The Commission of Colleges acted only in an advisory capacity; it lacked any legal or formal authority. Still, the New England Association believed that the commission's conclusions carried "great moral force" and, "so far as they have received attention from the colleges, have been promptly adopted."[103] Progress in unifying admission requirements and even in reducing differences in entrance exams had occurred, but the great diversity in the certificate system remained a pressing concern.

Most of New England's colleges readily embraced admission by certificate. It was an easier way to admit students, and, they thought, it brought the two levels into closer relations. The secondary schools, on the other hand, complained about the various forms the colleges required. A committee of New England professors studying the issue found that "in the form of the certificate great divergence appears."[104] This lack of uniformity posed a problem for the secondary schools, in the same way that the diversity in entrance requirements challenged them to prepare students in different ways for two or more colleges. Working again with the Commission of Colleges, the New England Association sought to eliminate the diversity in the certificate system and to improve it to ensure closer relations between the two levels.

To do so, the colleges and secondary schools returned to a proposal that Robert Keep, the principal of the Free Academy in Norwich, Connecticut, had advanced in the mid-1880s. He thought that a common board of examiners representing a consortium of six or eight New England colleges should inspect the secondary schools and determine whether they deserved the privilege of sending students to college on certificate. This proposal for regular inspections would bring the New England certificate system into line with the inspection model in place throughout much of the country, and it would ensure that the two educational levels had an avenue of fairly constant communication. In words that surely pleased Angell, Keep argued that such a board would open to the secondary school teachers "a wider horizon; would set clearly before them the type of scholarship and of mental training most valued in the college."[105] Students from schools that earned recognition then would be able to enroll in courses in any of the colleges participating in the board. In effect, the board would bring about greater uniformity in entrance requirements and eliminate the need for the traditional entrance examinations.

Angell, who pioneered the accreditation movement in Michigan, firmly believed that New England's certificate system could succeed if it took this step and established a robust system for inspecting and accrediting the secondary schools. Angell hoped that the region's colleges would unite to form committees that would divide the inspection work among all of the colleges in the region.[106] Such an approach would lessen the burden on any one school, while also building an effective inspection and accreditation program. Calls by Angell and Keep for a regional approach to accrediting schools were fifteen years early. In 1902, New England's colleges—spurred on by the New England Association and the Commission of Colleges—did form the New England College Entrance Certificate Board to act on and certify schools throughout the region, although it did so without providing for inspections. This board brought the colleges together, set standards for accrediting schools, and helped to create a uniformity in graduation rates and college entrance standards throughout the region.[107] Angell and Keep envisioned such a board in the 1880s, but New England's colleges had not been ready for such a shift from complete institutional autonomy.

# VI

Angell's model for accrediting schools and admitting students spread throughout the country, although not always in a form he would have liked. Colleges in New England and the South adapted it to their own particular circumstances, and some of New England's most prestigious universities refused to admit students by this method. Nonetheless, by the 1890s accreditation programs had become far more prevalent throughout the country than exams as a method for enrolling students in college. As a result, the secondary schools had a greater role than before in selecting who would go to college. With accreditation programs, the high school diploma and sometimes a recommendation from a principal or headmaster became the crucial credential that opened access to advanced study in many of the nation's colleges and universities. These programs helped to solidify the four-year high school as the dominant path to college, even as the secondary schools became an intermediate rung on the educational ladder.

The spread of accreditation programs and an emerging uniformity in admission requirements and exams reduced tensions between the two educational levels. Strides had been made in unifying entrance requirements and examinations, particularly in New England. In other sections of the country, accreditation programs made it easier for secondary schools to meet the demands of different colleges and universities, since many colleges began to accept students from high schools accredited by any college or university. The superintendent in Oshkosh, Wisconsin, could be happy that the road to college for his pupils was becoming less cumbersome. This growing consistency in standards and methods of admitting students was a remarkable achievement, given the absence of a regional or national board to compel such standardization.

The expansion of the accreditation system also made it easier for the high schools to prepare students for college and for life. This challenge of meeting the needs of students going on to college and of those stopping at high school was at the heart of attempts to bring higher and secondary education closer together, but addressing this challenge was time-consuming, as Eliot, Angell, Bancroft, and others were discovering. They had to deal with the crucial question of the purpose of secondary schools. Were these schools to prepare students for college? Or, did they have a larger goal of preparing students to be responsible, productive members of a democratic society? Was it possible for secondary schools to prepare students for both ends with the same curriculum? These were fundamental questions that prevented any easy approach to unifying secondary and higher education. As the next chapter describes, the secondary schools showed a significant resilience, strength, and assertiveness throughout this debate. They refused to bow to the mighty colossus that the universities envisioned themselves to be. By standing up to higher education, the secondary schools forced the colleges and universities to adapt to the lower schools.

# Chapter 4

# The Secondary Schools' Challenge to Higher Education and the Dominance of the Modern Subjects

## I

As a member of the school board for the Chicago Public Schools and president of the fledgling University of Chicago, William Rainey Harper was used to getting mail both supportive and critical of his educational policies. The letter he received from a self-styled "plain Citizen" in June 1898, however, was particularly blunt and critical. "There is a well defined and firmly fixed opinion in the minds of [a] large proportion of our very best people," it began, "that your main aim and plans are to run our public schools simply into a sort of annex or feeder to the University of Chicago. And your course so far in the Board of Education certainly is the grounds upon which they have been compelled to base it."[1] By the time he received this letter, Harper had detailed an ambitious proposal for the complete articulation of secondary education with his university, but, contrary to public perceptions, he never proposed turning Chicago's public high schools solely into annexes to his institution. "It must always be kept in mind," he said in 1892, "that the great majority of the constituency of the High School have no purpose to enter college. The work of the High School must therefore be adapted to the needs of those who regard this as their last student work."[2]

Still, Harper's plans and initiatives led many of the city's "very best people" to think otherwise. He promoted an ambitious scheme for inspecting and accrediting the lower schools and bringing them into a close relationship with his university. He did not stop with inspecting and accrediting secondary schools; he added teacher preparation programs and invited administrators and teachers to attend conferences and lectures on educational topics hosted by university professors. He also played an instrumental role in hiring a highly controversial superintendent for the Chicago Public Schools and then accepted a mayoral-appointed seat on the school board. Harper, of course, was not alone in inspecting schools, training teachers, or holding academic conferences with the high schools, but he packaged all of these initiatives into one coherent and bold program of secondary school–university articulation. Not surprisingly, then, some of Chicago's citizens responded angrily to what they perceived as the domination of the new University of Chicago.

In all of this—his open ambitions, extensive plans for closely articulating his institution with the secondary schools, and the opposition of many to those plans—Harper looked both forward in time and back to the late 1870s and 1880s. He foreshadowed some of the significant trends that the National Education Association's (NEA) Report of the Committee of Ten on Secondary School Studies and later committees would focus on in the mid-to-late 1890s. He looked back to a larger debate occurring in education and gaining momentum in the 1880s that questioned the real purpose of the secondary schools, both in relation to life and to higher education.

At the center of this debate was a series of crucial questions. Were the secondary schools essentially feeder schools to the nation's colleges and universities, or did they have an independent function entirely their own? This questioning, which went to the heart of education, only intensified as the public high schools in the late nineteenth century became the dominant form of education in a society undergoing far-reaching transformation. Were college preparatory courses the best preparation for students leaving high school and entering this changing social and economic landscape? What knowledge was most important for high school students to have, and did the knowledge of most worth differ if students were going to college or out into their communities as citizens of a Republic? Harper's critical correspondent essentially argued that the two functions could not be similar—that preparing for college and for life required different courses. As long as the gap between these two purposes remained, secondary schools had to maintain two courses of study, and this dual mission created sharp conflicts between secondary and higher education.

This chapter examines efforts by the secondary schools to reduce tensions between the two educational levels. It explores specifically the response of the secondary schools in the Midwest and throughout the country to Harper's ambitious plans and to the expectations of other universities. To do so, it focuses on the 1880s and early 1890s but also looks briefly at the late 1870s, when colleges first used their mounting expertise and authority to pressure the lower schools to prepare future collegians in a narrow range of subjects. Always autonomous to a certain degree, the secondary schools ultimately succeeded in broadening the focus of the universities to better reflect the work of the lower schools. They effectively pressured higher education to recognize as admission requirements the modern subjects commonly taught in the high schools and to create degree programs that aligned with these courses. These actions lessened the need for alternative courses of study to fulfill the dual responsibilities of the high schools. Such accomplishments benefited the middle class, whose children dominated the high schools and sought advanced study to earn college degrees and gain professional positions. Secondary schools thus were actors in their own right in the campaign to articulate education at the turn of the twentieth century, and their role in altering the requirements of many colleges and universities represented a striking achievement.

## II

Answering these elemental questions about the place of the secondary schools was crucial to building a stratified system of education, but the answers varied depending on the type of secondary school being considered. In New England, many of the preparatory schools—such as Phillips Andover and Exeter—had long traditions and histories of preparing students for college. These elite preparatory schools differed from most private academies, which functioned more along the lines of the public high schools in the Midwest. These public schools had a different place in the nation's loose structure of schools. Most of their students had no intention of furthering their education in college, and they needed an education that prepared them not for more school but for the duties of life. The debate over the purpose and role of the secondary schools for the most part, then, revolved around questions of the role of the public high schools and the private academies—as distinct from the preparatory schools—in an educational system.

Opinions varied widely in the late nineteenth century on the role of these public schools, and it was not unusual for people to change their minds and shift their opinions. Harvard's president, for example, altered his views over time. In the 1870s, Charles W. Eliot readily admitted that the high schools had a primary responsibility to prepare students for life.

"The first work of public schools, supported by local taxation, is not now to fit for college." As he explained in 1873, "Their work is to train their pupils in English, in mathematics, in classics a little, up to their seventeenth year."[3] He continued to hold this view for the next few years. "The public high schools," he declared, "have a different function [from preparing students for college], and the work of fitting a small proportion of their pupils for college, interferes with the discharge of their very important legitimate function." Given this view, Eliot doubted that the demanding preparation needed for college and the very different preparation needed for life could be combined. Yet, Eliot would not hold this view for long.[4]

Harvard, for the time being, was able to rely on its preparatory schools— including Exeter, Phillips Andover, Roxbury Latin, and Boston Latin—but institutions such as the University of Wisconsin had no such luxury. John Bascom, president the University of Wisconsin for much of the 1870s and 1880s, hoped to dismantle the university's preparatory department, and he had no choice but to depend on the high schools. As he asserted in his annual report in 1880, he believed that the state's high schools had a fundamental obligation to fully educate those students not going to college. This function, he suggested, was their primary work. Bascom nevertheless publicized the beneficial effects of the university on the lower schools. By asking the high schools to prepare some students for college, Bascom firmly believed that he and the university were elevating the standards of the lower schools for all students. He was impatient with the suggestion that the university diverted the schools from their more important work in fitting students for life. "If it be true," he began, "that the work done for the University diverts attention from the much more important work to be done for scholars who go no farther than the High School, we should accept the objection as a fatal one to any effort to unite the higher and lower grades of instruction by means of the High Schools." Such was not the case, as Bascom saw it. Rather, Bascom contended, the university strengthened the secondary schools and gave them a more productive and rewarding purpose by inducing them to look beyond their narrow work "to the great stores of knowledge" that the universities possessed. In other words, he began to assert, as Eliot later argued, that preparation for college and for life could be similar. However, Bascom implied, both pursuits needed to rest on a college preparatory course of study.[5]

Bascom's words were not fully persuasive, and George Peckham of the Milwaukee High School challenged his report. He regarded the University of Wisconsin as a domineering institution that tried to deflect the secondary schools from educating the vast majority of students who had no desire to further their education. "So far as my observation extends," he wrote in 1881 in the *Wisconsin Journal of Education*, "the dominant sentiment on this question is the exact reverse of the one suggested by President Bascom. That as a matter of fact the needs of the large number of students who finish their education at the high school are sacrificed to the needs of the small number preparing to enter college." Sacrificing the needs of most high school students for the "three or four fitting for college" was an abomination, Peckham reasoned, since it meant transferring teachers and money from "political economy, geology, physiology, and hygiene" to classical languages and other preparatory subjects. The college preparatory course was perfectly fine, he asserted, when topped off by four years in college. For those students ending their education with the high school, such a course "was exceedingly one-sided."[6] He hoped all students who desired to would continue their education in college, but he refused to concentrate on this small number to the exclusion of the other students. Peckham clearly sided with the majority of his students against the university's president.

Most of his colleagues agreed with him. The high schools had their own vision for the subjects they needed to offer their students, driven, in part, by the needs and desires of the surrounding communities. To many of Wisconsin's citizens, especially in rural areas, the university in Madison—or in Ann Arbor, Berkeley, or Cambridge, for that matter—was little more than a glimmer in the distance. These local communities, to the degree that they supported their high schools—and some support was precarious— had little appreciation for the classical languages and the traditional college preparatory curriculum. The public, whose support was vital to the success of the high schools, ensured that these secondary schools developed modern courses as preparation for life and did not focus exclusively or even primarily on classical courses as preparation for college. Yet, the high schools understood that they could not neglect the growing number of secondary students seeking to go to college in the late nineteenth century. Many of these college-bound students came from the middle class, and, while middle-class parents often supported articulation and increasingly wanted their children to go to college, they valued the modern subjects over the classical courses, and the access such practical subjects could provide to the professions. Caught between public demands and college expectations, the high schools struggled to find a way to meet both needs.[7]

What was never particularly clear in these discussions of the role of the secondary schools was what preparation for life or for college really meant. The high schools knew that they could not focus on Latin and Greek, even though the colleges continued to assert that such classical preparation and a liberal education provided the best way to know and understand the world. For the colleges, liberal education unfolded to young minds the richness and knowledge of centuries of life. Greek and Latin, the ancient languages needed to unlock this knowledge, represented the foundation of true education. Over time, the colleges came to accept that the modern subjects could afford access to this culture, but the classical languages remained for them the key to a strong education. The secondary schools, in contrast, shunned the overriding focus on classical preparation, especially Greek, to offer a broader array of modern courses—English, modern languages, algebra, science, and history—that would help students navigate their way through society. Modern foreign languages, for example, were practical since the nation traded with French, German, and Spanish-speaking peoples, and scientific study proved helpful in exploring the nation's abundant natural resources. Although high schools never saw their purpose as training students for specific jobs—a focus on such vocational education would come later in the twentieth century—they also concentrated on courses that gave students some skills useful in occupations, such as commercial arithmetic, bookkeeping, and domestic arts.[8]

These two approaches—broadly labeled liberal and practical education—were at the heart of the debate and tension between the secondary schools and the colleges and universities. To be sure, these labels were misleading. Not all educators at the secondary school level fully agreed on the aims of the public schools or what preparation for life meant. Some saw their duty as preparing students to be citizens, while others claimed that the public high schools had an obligation to prepare students for business and work. Still other teachers reflected college ideals of liberal education, although they were careful to insist that the modern subjects could provide such an education. These were ambiguous labels that never neatly represented a wide array of views, but they did highlight the tension between what the high schools saw as their main purpose and what the universities increasingly needed from them.

The superintendent in Ann Arbor, Michigan—an exceptional high school in part because of its proximity to the state university—keenly felt the tension between preparation for college and for life. The growing number of students seeking to prepare for the university at the Ann Arbor High School, he knew as early as the 1870s, was "rapidly changing the character of the High School from a department proper of our public schools to a school of preparation for college." One course of study for all of the students in the school was not a feasible option. "In order to maintain this double character" in preparing students for life and for college, he claimed, "several co-ordinate courses of study must be kept up; for what is considered best for those who finish their studies with the High School, in no way prepares them for University work."[9] He was not in a position to emphasize the non-preparatory courses at the expense of the preparatory subjects, since his school was located in the same town as one of the West's strongest universities and was seen throughout the state as a feeder to that institution. Being a relatively large school, however, Ann Arbor had more resources than smaller schools and could offer dual courses of study that led both to college and out into life.

Administrators from other Michigan schools, even the smaller, rural schools, accepted that they had an obligation to prepare students for college. Michigan, after all, lacked a strong tradition of private preparatory schools, and the university, without its own preparatory department, depended on close relations with the state's high schools to funnel students to its classrooms. The state's principals and superintendents recognized their relationship to the university, even though they made it clear that their schools had a purpose beyond college preparation. School officials in Mt. Clemens asserted that the primary purpose of their school was to take students from the lower grades and expand on that education, "so as to prepare the pupil, as he may elect, for admission into the ordinary branches of business, or into the Normal School or the University." The mission of this school reflected the dual pressures on the high schools. "The high school has a two-fold nature; first, to make the scholar a useful citizen when he goes forth from its walls; and, second, to qualify him for admission to the University, if he shall desire to go there."[10] Battle Creek's administrators sympathized with the Mt. Clemens school and valued the high school's role in preparing some students for college, but the school's superintendent left no doubt that he thought an education "best adapted to the common experience of life" was the "chief end of the High School." He refused to argue that college preparation should not be a function of the high school, but he wanted to make it abundantly clear "that very many who enter [the high school] do not expect to go beyond it, but desire a thorough, practical, academic education, which shall be as complete as may be in itself."[11]

Michigan's schools were not alone in having to balance competing demands. The purpose of the schools, one of Wisconsin's county superintendents declared in the early 1880s, "should be to lay the foundations of knowledge merely, to prepare generally for all pursuits, specially for none." He continued to make a utilitarian argument that "the controlling fact in the course of study should be the greatest good of the many pupils, and the many will pass from the high school into the world." Preparation for college, he conceded, could be a part of the function of the high school, but it was to be a "secondary function" that "must neither destroy nor control the primary functions." He trusted that the better schools in the state could take on this secondary function without any "material loss to other interests."[12] The *Wisconsin Journal of Education* jumped into the discussion and agreed with the county superintendent. The primary function of the high school was to fit the needs of "the great majority of the students [who] wish to prepare themselves for business life." Larger schools, the editorial commented, could provide a variety of courses but smaller schools had to meet the needs of their communities.[13] These needs focused on the demands of life rather than on the demands of the colleges.

Other educators were not inclined to be so supportive of the needs of the universities and colleges. The superintendent in Port Huron, Michigan, expressed his point of view adamantly. "There is a great cry among our educationists as to how we shall model the High School to suit the requirements of our University," he maintained in 1875. Rather than taking such action, he proposed a different step. "I think that the question should rather be how the University can be modeled to suit the requirements of the High School. To nineteen out of every twenty students," he argued, "the High School is their University. They go out into the world without any further preparation, and I think that our energies ought to be devoted to the general education of the nineteen rather than the special preparation of the twentieth." If the university wanted students from Port Huron, the local superintendent argued forcefully, it would have to accept them as the high school prepared them.[14]

The superintendent of the Springfield, Missouri, schools held similarly strong views. He steadfastly refused to see any connection between the secondary schools and the colleges and universities. "The high school should have nothing whatever to do with the college," he announced in 1885, and it should design its course of study without considering the needs of the colleges and universities. "Possibly, throughout the country, not one high-school graduate in fifty will ever enter a college, and we do not desire to shape a course of study to accommodate this one fiftieth part of the students. It would not be wise." Unwilling to countenance college preparation for the majority of his students who could not afford a college education, this superintendent stood firmly against those university presidents who wanted college preparation to be a significant part of the high schools' work. "The high school is really the poor man's college," he believed, and he refused to let the needs of the colleges compromise that laudable goal.[15]

In declining to sanction college preparation, superintendents in Springfield and Port Huron were in the minority. Many of their colleagues around the country accepted that the schools had a responsibility to prepare students for college. This responsibility gained importance as American society changed and demanded well-educated leaders able to take positions in industry, government, and business. For many of these new positions, a high school education no longer provided the necessary credential. In Michigan, Wisconsin, and other states, administrators recognized something that Eliot took longer to comprehend. Students who wanted to go to college and have access to these new positions had few options outside of the public high schools. The steep costs of private preparatory academies prevented many families from taking advantage of the opportunity to go to college. Eliot came to understand this reality by the mid-1880s, and his views on the public high schools shifted accordingly. "Since the high school supplies the only means by which parents who cannot meet the charges of private schools or academies can get their children prepared for college," he said in 1885, "it is much to be regretted that the number of students who make their way to college from high schools is so small absolutely, and so small relatively to the number of students in these colleges." In the 1870s, he promoted little if any role for the public high schools in preparing students for college. Now, more than ever, he looked to the high schools as an important avenue to higher education for some of the nation's young.[16] Harvard's need for students and the increasing dominance of the high schools likely contributed to Eliot's shifting attitude.

As he changed his position on the public high schools, Eliot found himself struggling with significant challenges. Even though the number of students in the public high schools increased considerably between 1866 and 1885, the number of students entering Harvard from the typical high school was minimal, and few students from public schools in Massachusetts even enrolled in college.[17] "Counting both boys and girls," Eliot concluded, "the high schools did not, on the average, send one pupil apiece to college in 1884, and just about one in a hundred of all the pupils in all the schools got to college in that year." Most of the schools in Massachusetts were too small to provide a strong preparatory course, and devoting limited resources to such a course was not tenable, especially since few students left those schools for advanced study. Asking these schools to focus all of their energies and resources on meeting college entrance requirements, Eliot concluded, was not possible. "It is impossible for such feeble schools to maintain a course of study which will regularly prepare pupils for college," he asserted. At least three teachers were necessary, Eliot believed, to offer a strong preparatory course, and he knew that more than two-thirds of the state's high schools lacked the proper number of teachers. "The high school is obliged to provide, as well as possible, for that great and increasing majority of its pupils whose education is not to be prolonged beyond the school, and can have, as a rule, but very limited resources to be used for the exclusive benefit of the small minority who hope to go to college."[18]

Eliot was coming to understand the problems posed by the conflicting demands of the colleges and the historic mission of the public schools, and he recognized the challenges that this lack of connection between the high schools and the colleges posed for education and for society. "The rigidity of the college requirements for admission, and the ambition of the colleges to advance their standards on the one hand, and on the other the legitimate demands of that great part of its pupils who have no use for Latin and Greek," he argued in 1885, "make the position of the high school more and more precarious, and its work of preparation for college less and less adequate, and therefore make it harder and harder for Massachusetts farmers, mechanics, operatives, clerks, tradesmen, and professional men of small income, to send children to college." Eliot lamented that "contrary to the interests of the Commonwealth, and of society at large, these classes are being measurably cut off from the colleges."[19]

Even though his rhetoric at times emphasized the importance of education in a democracy, Eliot never expected nor wanted all young Americans to attend school and graduate from college. He doubted that education and schools could equalize men and create a classless society. Nonetheless, he understood that the inability to resolve the tension between the dual functions of the high schools would have lasting implications for society. "For lack of adequate connection between the high schools and the colleges, the way to the learned professions and the best posts in all the highly organized industries," he concluded in 1885, "is being obstructed for large numbers of promising young men."[20] He and others needed to find a way to bridge the gap between the high schools and the colleges so that more students could enter college and, through that training, move into the professions needed by a changing society.

Eliot's dawning awareness of the importance of the high schools to society and the bind they were in was crucial to addressing the tension between preparation for college and preparation for life that Bascom in Wisconsin and later leaders such as Harper in Chicago worried about. "Broken or obstructed connection between the public secondary schools and the colleges," Eliot maintained in 1885, "is an evil which every friend of education must wish to cure."[21] The larger and better equipped high schools attempted to overcome this problem and meet both functions by creating two courses of study, one that led up to the university, the other to life. The smaller, typical high schools—often situated in rural areas and far from larger cities with thriving high schools—lacked the resources and public support to offer dual courses of study. At best these smaller schools offered a basic course that extended the elementary program by a few years. With the local high school providing only a limited secondary course and the better high schools located at a distance, students in rural areas found they had limited opportunities for the secondary preparation needed for advanced study at a university. James B. Angell learned through the accreditation program in Michigan that the public would not allow the high schools to develop a single focus on college preparatory work. Likewise, Eliot came to understand and accept that the high schools were not capable of redesigning their entire curriculum to meet college entrance requirements. The secondary schools were willing to make adjustments and reach up to the colleges, but they were not going to radically redefine their work in line with college demands.

# III

To work with the secondary schools and relieve the pressures on them, and to encourage more students to enter college, Eliot and his colleagues at other universities began to expand their admission and degree requirements to fit what the high schools were offering. "The remedy for the serious evil which thus results from the diverging aims of the high school and the college is to be found," Eliot claimed in 1885, "in the introduction into college requirements for admission of reasonably wide options, so that some course or courses of study which will admit to colleges may be brought almost to coincide with a substantial high-school course of study, laid out primarily for youth who are not going to college." Eliot was sounding like a high school administrator. In creating this ideal course, he turned to many of the subjects already present in the typical high school course— "English language and literature, mathematics through trigonometry, drawing, the history of England and the United States, physics, chemistry, botany, and zoology taught with instruments and objects in hand."[22] For good measure, he added some easy Latin and French prose.

As Harvard and other universities broadened their requirements, they included those subjects that comprised a good public high school curriculum. They were not yet ready to add more "practical" or vocational courses— manual training, commercial arithmetic, drawing, or domestic arts, for example—to college admission subjects, but they were reaching out to the high schools and making some adjustments in line with secondary school subjects. Eliot proposed that the universities merely accept what the high schools had already been teaching for most of the nineteenth century. These modern subjects were becoming college entrance courses, with the potential then of easing the tension between the two educational levels and reducing the gap separating them. Reflecting a position that he would continue to develop in the 1890s, Eliot trusted that this "ideal high-school course would be just as useful,—however utility be defined,—to the future candidates for college as to the mass of the pupils." Eliot argued that such a course would ensure that the high schools did a better job of preparing students for college, without forcing them to adopt classes that would be useless for noncollege-bound students.[23]

Likewise, Harper found that his university had to adapt to the needs of the high schools, specifically the Chicago high schools that some residents claimed he was trying to turn into preparatory schools. "Several subjects commonly taught in the high schools of Chicago and in secondary schools elsewhere had not received recognition at the hands of the University," he wrote in 1898 as part of his review of the first six years of the university's work. As a result, "a pupil in these schools who had fulfilled with credit the requirements of the normal four-years' curriculum" was often unable to enroll in the university. Harper attempted to resolve this unfortunate situation by accepting additional subjects—those taught in the Chicago public schools—as entrance requirements. In 1896, to address the needs of the local schools, the University of Chicago added "elementary and advanced United States and English History and Civil Government" to its list of admission subjects.[24] Nonetheless, Augustus F. Nightingale, superintendent of high schools in Chicago, felt that the university needed to adopt even broader admission standards to make it accessible to any student who had completed a strong four-year high school course of study.[25]

As countless high school educators pointed out, the traditional college preparatory curriculum made it difficult to meet the needs of college-bound students and those with no further education in their futures. In 1885, when Eliot hoped to change Harvard's standards, his institution— and most other universities—required future students in the classical program to study Greek, Latin, French or German, and "the elements of mathematics and physics, a little ancient history, and something of English literature."[26] Although the focus remained squarely on the classical subjects, the addition of some modern classes represented a significant advance from the situation throughout much of the nineteenth century. On the eve of the Civil War, the nation's colleges generally required only a few subjects for admission: Latin, Greek, mathematics, some logic and moral philosophy, and a little physics and astronomy. Following the war, they gradually added, as Eliot was doing, English grammar and composition, algebra and geometry, geography, history, and more sciences.[27]

At the same time that the universities began to place more of the modern subjects in their entrance requirements, they also added alternative degree programs for students who prepared primarily in the modern subjects. Not wanting to sully the classical bachelor of arts degree with a strong focus on modern subjects, these institutions created parallel degree courses that, although less prestigious, allowed schools to prepare students for higher education without having to concentrate limited resources on classical training. These new programs first emerged on college campuses in the

1850s, with Brown offering the bachelor of philosophy degree in 1851. Harvard in 1847 opened the Lawrence Scientific School—a new school that by design separated its degree program from the classical work of the college—and began offering the bachelor of science degree in 1851. Yale likewise offered the bachelor of philosophy degree in 1852 through the Sheffield Scientific School, and Michigan added the bachelor of science degree in 1853. These universities were only the first to add the new courses.

After the 1870s, these alternative courses proliferated on other campuses, and even more degree programs evolved. The University of Wisconsin first offered a modern classical course in 1876, an English course in 1887, and even a civic–historical course in 1893. The number of alternative degrees multiplied on American campuses in the last half of the nineteenth century—some schools eventually offered nine different degrees—but by 1890, most colleges enrolled students in one of three or four programs. The traditional classical degree remained the most esteemed but, in relation to the other degrees, it gradually lost students who looked to the new courses as better routes from high schools to jobs. A semi-classical course—either the bachelor of philosophy or the bachelor of letters—aligned more closely with the modern subjects in the typical public high school. The bachelor of science course stood apart as having almost no grounding in the classical subjects, except that over half of the schools offering this degree required some preparation in Latin.[28] Angell trusted that all of the high schools in Michigan would be able to prepare students to enter the university and study for the bachelor of science degree. "There is," he declared in 1875, "no respectable High School in the State, which cannot do preparatory work, that the University with perhaps some unessential modification of its present requirements, can properly accept as suitable for its scientific courses."[29]

Angell must have concluded that the science degree was not connecting the high schools and the university to the extent he wanted, because in 1878 his faculty developed the bachelor of letters degree to align directly with the "English" course of the high schools in the state, a course that emphasized English literature, history, government, and modern languages rather than the classical subjects. According to Angell, the faculty felt that, since most of the high schools offered a thorough English course that surpassed the classical course in depth of preparation, the university had an obligation to create a corresponding degree. In the absence of a strong college degree program that aligned with one of the state's most prevalent high school courses of study, the gap between secondary and higher education remained unclosed. Angell and his faculty asked, "Ought [the university] not to try without sacrificing the interests of good scholarship and sound culture, to bring itself into some harmonious and useful relation with that large number of High Schools which provide no classical course, but do provide a thorough English course of education which may form a suitable preparation for some scholarly course of training here?" They answered "yes," and to better meet the needs of the public high schools, the University of Michigan created another alternative degree program. The university succeeded in creating a new course that aligned with the high schools, and it was one that ultimately brought more students into the university. "So far as numbers indicate," Angell pronounced in 1879, "we have certainly good reason to be satisfied with the response which has been made to our proposition to make our instruction more attractive and useful. The number of students in the Literary Department is increased by about *twenty per cent.*"[30]

Higher education better met the expectations of the secondary schools through these new degrees, but the desires of the high schools alone did not lead to changes in admission subjects and degree programs. The universities also responded to the changing needs of society and to the advances made through scientific and technological innovations. The sciences, modern languages, and history, for example, were more relevant to the needs of an industrializing society than Greek or Latin. These newer subjects would lead to engineers who could build skyscrapers, scientists who could study diseases, and historians (as Frederick Jackson Turner famously argued) who could instill the values of the Republic in students and prepare them for democratic citizenship. Scientific and engineering courses, in particular, gained strong currency in this context, and colleges and universities began to require more preparation in the sciences for entering students. New academic fields also proliferated in the late nineteenth century, as professors engaged in research in modern subjects and devoted time to new areas of expertise. These professors had an interest in adding these subjects to the admission requirements as a way of bringing in well-prepared students capable of work in these areas. By developing strong research agendas in modern subjects, university professors contributed to the legitimacy of the modern subjects in higher education. The needs of society and the emergence of universities came together with high schools and their traditional focus on the modern subjects to create a closer connection between the courses of study in higher and secondary education.[31]

Because the new nonclassical degree programs generally omitted Greek in lieu of the modern subjects, they opened up a college education to a greater number of students than did the standard classical course. These degrees recognized that the high schools, constrained by local needs and expectations, offered a richer program in the modern subjects than in the classical subjects of Latin and Greek. Prior to the emergence of these degrees, rural students had few opportunities to prepare for college. Secondary schools rarely could afford teachers in Greek and Latin, and the local communities usually had no interest in providing such courses. Ambitious students often had to move away to a larger town or enroll in a private academy to prepare for college or they had to forego any interest in college at all. By adding new degrees, the colleges and universities opened up the possibility of further education to these students. These degrees improved the options available for students seeking a college education, and they made it easier for students from the public high schools to enroll in college, graduate, and enter a workplace transformed by science, technology, and corporations.[32]

While helping students without classical preparation to attend college, however, the creation of alternative courses and the addition of typical high school subjects to college entrance requirements did little to unify the diverse admission standards throughout the country. Indeed, since each school set its own entrance requirements for these new courses, the proliferation of degrees and subjects actually created greater diversity in entrance requirements and, thus, challenges for the secondary schools, even though these additions provided more opportunities for students. Schools prepared students not just for a number of different colleges but also for two or three degrees within a single college. The new degrees brought the colleges more into line with the typical high school—an outcome greatly desired—while at the same time increasing the diversity of preparation for those schools that groomed students for different degrees and colleges.[33]

There were other problems with the proliferation of degrees. Not only did they increase the diversity of requirements for admission, they also often maintained lower standards relative to the bachelor of arts degree. Some universities went as far as enrolling students in these degree programs before the students had completed a full secondary school course, which predictably led the secondary schools to complain that higher education had overstepped its bounds. The principal in Middleburgh, New York, offered a typical protest. "The new courses," R. S. Keyser declared in 1887, "are mostly designed to attract students by offering them admission to college upon easier terms. Instead of completing the work of the secondary school, the college thus comes into direct competition with it."[34] The bachelor of science degree was especially weak. An article in the *Academy* in 1887 declared, "We simply ask that the scientific course in college should not be brought into competition with the proper work of secondary schools, and that the requirements of the scientific course should be such as to require the same age and the same mental maturity on the part of students that is required in the classical course."[35] Rather than strengthening the alignment between higher and secondary education, the proliferation of degrees, especially the bachelor of science course, sometimes weakened that connection and brought the two levels into competition.

Michigan made a series of strategic decisions in the late 1880s that strengthened its degree programs and helped to reduce tensions with the high schools. The university's faculty in early 1888 proposed some basic alterations in admission requirements to bring about greater equivalence among degree programs in the amount of preparation and study required. Although the recommended changes did not require all students to study the same subjects or topics, they helped to establish the core of a unified course of study. This step worked to the advantage of the high schools, which had faced the challenge of meeting different demands and expectations for the various degrees. All of the degree programs, according to the proposed changes, required students to prepare fully in Greek and Latin (with German and French as alternatives), mathematics, physical and biological sciences, English language and literature, and history. These subjects formed the principal admission requirements, with other subjects added and more or less work expected in some depending on the degree program a student ultimately entered.[36]

The faculty recommended trying this program for three years in the high schools, analyzing the feedback from the schools, and then deciding whether to make additional changes or continue with the new requirements.[37] The high schools did not wait to offer their feedback. Within a month of these new proposals being discussed in university meetings, teachers in the Ann Arbor high school protested the increased work in physics required by the changes. While the increase sought only to equalize the preparation in physics required for all of the courses, it specifically affected the bachelor of arts and bachelor of philosophy courses, and the teachers asked for some reduction in the mathematics requirements to make room for the new physics work. Too much else was expected, they complained in 1888, for these degree courses to absorb additional preparation in physics. Ann Arbor's faculty likely had little opposition to the establishment of core classes for all degree programs. What they protested was the addition of subjects to some of the preparatory programs without a corresponding reduction in other classes. Continually adding courses without subtracting others overloaded the school days, they maintained, and did little to alleviate the pressures on the high schools.[38] Whether it satisfied the request of the teachers or not, the university did lower its mathematics requirements before adopting the new standards.[39]

The University of Wisconsin went beyond Michigan's efforts in equalizing its courses and in meeting the needs of the secondary schools. As the president proposed in 1891, he wanted his institution's requirements to align with the courses of study and standards set by the state superintendent. The president embraced these courses, in part, because they all required a core set of subjects and expected an equivalence of work. Wisconsin also provided some flexibility so that high schools could adapt the courses to meet local circumstances and needs. In a letter to the state's principals, Wisconsin's president asked whether the new requirements met their approval, and he sought their feedback. He was loath to make any changes without first consulting the high schools, even though the changes proposed were in line with the courses developed by the state superintendent.[40] Wisconsin implemented the new standards in 1892, with the expressed hope of bringing the "preparation for all courses up to an essential equality." Moreover, "the requirements were so arranged," the president argued, "as to bring the University into more intimate and formal relationship with the state school system." He concluded by declaring that the high schools "most generously" complied with the new standards.[41]

In contrast, Eliot at Harvard avoided some of the conflicts caused by the differing standards of the new degrees by simply not rushing to offer additional programs and courses. Harvard did open the scientific school, which, to Eliot's chagrin, maintained relatively low standards, a situation he labored to rectify in his four-decades-long tenure as president. But this degree—housed in a separate school—represented the only variation at Harvard from the traditional bachelor of arts degree. Instead of adding degrees, Eliot tweaked the bachelor of arts program in the mid-1880s and sought to make this degree more appealing to students from the public high schools. Not wanting to cut off access to students in an expanding middle class from a Harvard education, Eliot broadened his entrance requirements and, on paper anyway, made it easier for some students from small, rural high schools to enter his university and earn the classical degree.[42]

Eliot's new proposal essentially promoted an elective system for the high schools by allowing a greater range of modern subjects as admission requirements. This change provided enough flexibility in course selection and enabled most high schools to meet the needs of all students, regardless of their future destinations. Under the new standards, all those admitted to Harvard earned the bachelor of arts degree—unless they enrolled in the Lawrence Scientific School—but they chose from a number of different subjects to meet the entrance requirements. To make Harvard even more appealing, Eliot sacrificed one of the crucial admission components of a classical education. This bastion of elite education no longer absolutely required Greek for admission in the 1880s; instead, students could substitute courses in mathematics and the physical sciences. Eliot was trying to make it easier for farmers, mechanics, and "professional men of small income" to send their children to Harvard.[43] Combined with the trend toward core subjects for all degree programs in other universities, Harvard's actions opened up the possibility that high school students could prepare for various degrees without having to take significantly different courses. Harvard also made it easier for schools to prepare students for life and for college with one unified course of study and without Greek. With these modifications, Eliot signaled his willingness to work with the high schools in resolving the tensions they faced in preparing for college and for life.[44]

Keyser, the principal in Middleburgh, New York, applauded Eliot's new openness to a broader range of admission subjects. "Harvard realizes that conditions have changed, and that if she desires a wider constituency, she must look to the constantly increasing list of public schools," he declared approvingly in 1887. "Her new requirements for admission offer extra inducements to public schools to send up their pupils, permitting the substitution of other high school subjects for Latin and Greek to almost any extent." Keyser clearly understood the changing nature of education and the increasingly prominent place of high schools by the 1880s and 1890s. He encouraged other schools to follow Harvard's lead and lessen their reliance on the classical languages. "In the case of the classics, it is well to consider whether there is not need to throw overboard some of the cargo to save the ship," he argued. "The simplest solution of the question, which would bring the classical course more into harmony with the actual educational work of the country, would be to decrease somewhat the amount of Latin and Greek required for admission to college, as Harvard has already done."[45] For this principal, anyway, Harvard had made important strides in adapting its policies to the reality of the secondary schools.

Yet, he had been overly optimistic about Harvard's lead. This elite institution had not gone as far as Eliot had implied. Although he made it possible to enter the university and study for the classical bachelor of arts degree without preparation in Greek, Eliot made the alternative requirements in mathematics and physical science more difficult. He was willing to expand the road leading to college, but he wanted the new avenue to be as rugged as the classical road.[46] Equivalence for him meant making all subjects equal in standards and rigor. Modern subjects were not given a free pass but had to match the academic excellence maintained in the classical subjects. Eliot indicated a willingness to work with the high schools and to widen the road to college, but the passage remained difficult.

The nation's colleges and universities, therefore, had made important changes in admission requirements and degree programs to bring about closer articulation with the secondary schools, but complete alignment was not yet in sight. The high schools still felt trapped by the demands of the colleges and the needs of life. William Torrey Harris, the U.S. commissioner of education, made it clear in 1891 that the tension between the two functions continued to challenge secondary schools. "In the schools of the United States there prevail two different ideals of the course of study; the one originating with the directors of higher education and the other a growth from the common elementary school." Harris, the former superintendent of schools in St. Louis, said that these two ideals were not compatible and clashed "in quite important particulars." The heart of the issue, for Harris and many others, was that "the common-school course of study, as it appears in the elementary school and in the public high school which gives secondary instruction, does not shape itself so as to fit its pupils for entrance to the colleges."[47]

As Harper's critics in Chicago made clear in the mid-1890s, the debate between preparation for college and preparation for life had not ended, despite progress in aligning university courses to the secondary schools. The debate remained tense. In 1891, before Harper opened the doors to his university, Cecil Bancroft moved beyond Eliot's tentative steps at Harvard and offered what he hoped would be a viable solution. As principal of Phillips Academy, Bancroft proposed a unified course of study that sought to make preparation for life and for college the same thing. "How can courses be devised," he asked, "which shall meet at once the wants of pupils soon to be plunged into the distractions and responsibilities of their vocations, and of that other and smaller number who have before them the prospect of long courses of further training?"[48] Bancroft took steps to unite the dual purposes of the secondary schools, but his proposal likely would not have pleased Harper's opponents, since Bancroft's focus was on preparing more students for college.

The question for him was not how to create a unified course of study so that the secondary schools did not neglect the majority of its students— those who would never go to college—but how to devise such a course that would help these students while also encouraging an ever greater number of them to press on to college and then become society's leaders and experts. "Our secondary schools ought to embrace such subjects, and prosecute them in such a way, that boys and girls in country towns, as well as in large cities, shall have the opportunity to reach the college and the university." Bancroft worried that separate courses of study forced students to choose their future paths—whether to college, scientific school, or life—at too early an age. He wanted to postpone this decision until late in the secondary school so that students might have a chance to feel the vivifying effects of education and consider the possibility of advanced study. "All the roads in Italy led to Rome," he explained by way of analogy, "and many a traveller setting out for a brief journey found himself lured on by the circumstance of being fairly on the way, till at last he came to the seven hills and beheld the glories of the eternal city." In the same way, "it is no small service to our youth to keep open before them, till the last moment, possibilities of the best education our civilization offers. This is a noble function of the secondary school" and one that the country absolutely needed.[49]

Creating a unified course of study provided an avenue for strengthening the connection between higher and secondary education, and, as Bancroft wished, the proliferation of degrees and subjects constituted the raw material for just such a plan. The abundance of this raw material, however, made the development of a uniform course complicated. Who would decide what students needed to know? Bancroft signaled his willingness to think anew about the course of study. "We have fairly well emancipated ourselves," he concluded, "from the old tradition that certain subjects are liberal and all others are not, that the humanities alone are capable of imparting that tone and temper of mind which constitute true culture." He argued that "the passionate love of the search for truth, may be reached through sciences, through history, through modern literatures, as truly as through the conventional Latin, Hebrew, and Greek which constituted the staple studies of our early colleges." He was even willing to consider that manual subjects could lead to truth if pursued in a "liberal spirit." Bancroft may have freed himself from the hold of the classics, but others had not yet been able to make that shift. Rather than resolving the issue and providing the definitive answer, Bancroft only highlighted a significant challenge to those reformers who wished to articulate the two educational levels.[50]

The lack of agreement on what constituted a proper course of study, of what knowledge was of most worth to an educated person, hindered attempts to align higher and secondary education. Bancroft proposed some ideas and Eliot was willing to relinquish the hold that Greek had on the core of what it meant to be an educated person, but these actions never resolved the issue. Indeed, there seemed to be a curious lack of discussion of this topic in the debate over the relationship between the two educational levels. True, Bancroft and Eliot made some suggestions, and Harris was always willing to expound on his theories of what constituted a true education, but most discussions stepped gingerly around the issue. Articulation proponents did not devote discussions to theories of the educated person or to the knowledge that educated people needed to have at their command. In the same way that the process of strengthening the connection between the two levels lurched toward some conclusion without much coordination, the debate over what constituted a true education progressed in a haphazard way without any clear guidance or control. Even most famous education report of its time, which came out a few years after Bancroft's passionate plea, avoided the topic. Eliot's report from the Committee of Ten on Secondary School Studies, issued in 1893, simply accepted and dealt with the subjects that most schools had already been teaching. The answer to what constituted a proper education was whatever tradition, societal needs, and pressure from the universities and schools dictated. The result was a less than coherent grouping of subjects, courses, and degrees, perhaps because the needs of society were varied and constantly shifting.

High school teachers also lacked consensus on this issue. Some agreed with Bancroft and ardently believed that the best preparation for college was also the best preparation for life. "If the college work is in reality the crown of a sound education, then all the work preparatory to it must also be sound, and that a boy or girl prepared for the colleges will also have the best preparation that can be given him, thus far, for life," argued Isaac Thomas, the principal of Hillhouse High School in New Haven, Connecticut, in 1892.[51] For him, the college preparatory course provided the best foundation for life, rather than the needs of life forming the best preparation for college. Others saw the needs of the college-bound as separate and distinct from those who were going out into life with no further education. Thomas's counterpart at the Free Academy in Norwich, Connecticut, countered that pupils who entered the noncollege course "are to have only these four years of schooling." Consequently, the secondary school course, he continued,

> ought to give them, though not in the same way, the most important thing to be got from a college course: namely, an idea of what liberal culture means, a love for it and a desire to get as much of it as they can; an outlook on life which will make their lives something more than a humdrum round of toil relieved only by common-place enjoyment.

There were similarities between preparation for college and for life, he maintained, but the needs of the two were not the same.[52] Even with the addition of new courses and modern subjects to the college curriculum, the debate over preparation for life and college clearly had not ended in the 1890s, when the University of Chicago opened its doors and citizens started to complain that Harper's grand plans meant making the high schools preparatory departments of the university.

# IV

The high schools, and the public that supported them, continued to protest what they perceived to be university attempts to mold the secondary schools into classical preparatory academies. But observers in the 1890s found a stronger and more clearly articulated system of education than they had seen in the early 1870s. A more prominent place for modern subjects in college admission requirements, the emergence of alternative degrees, and a push toward equivalence among modern and classical subjects, which would intensify in the future, contributed to an improved connection between the higher and secondary branches of education in the 1880s and 1890s. Additionally, the expansion of accreditation programs throughout the country eased the pressure on secondary schools to prepare students for different colleges. Eliot and the Committee of Ten now attempted to build on these improvements, to formalize them in an official report, and to promote them as a national and rational answer to the challenges of building a standardized system of education across the states and regions.

Eliot's task was significant. The colleges and universities were gradually accepting the modern subjects as admission requirements and developing alternative courses in an effort to better align with the secondary schools. These actions, however, did not mean that uniformity from college to college always existed. The colleges generally agreed on the subjects to be included in the admission requirements, but they were not yet always in agreement—especially across regions—on the specific topics and aspects to be required in each of the subjects. The English requirements in one college, for example, did not necessarily match the English requirements elsewhere. The work of the Commission of Colleges in New England on Admission Examinations had helped to bring about some uniformity in a handful of entrance subjects. The spread of the accreditation system similarly had created some standardization in the Midwest and West. Eliot's committee now sought to nudge these tentative steps toward fruition on a national level.

The members of Eliot's committee also had to wrestle with a number of other pressing issues that remained unresolved. It was not clear whether the different degrees that colleges offered had any equivalence in breadth of study and admission requirements. Neither higher education nor secondary education wanted one collegiate course to be markedly easier than the others, require little if any secondary education, and lead students to leave school early. Additionally, the two levels continued to debate whether preparing for college and for life required dual courses of study or could be met with one program. Educators often used ringing, inspiring language in reference to a unified course—and it was a vivid response to a complex problem—but the lofty language seldom was soiled by the gritty reality of hashing out and defining such a course. A more prominent place for modern subjects in admission requirements had eased the pressures on the secondary schools, but major representatives of the two levels had not agreed on whether a unified course of study could meet mutual ends. They had difficulty in reaching a conclusion, in part, because they disagreed on what preparation for life or for college meant. What it meant to be an educated person and what subjects were of most worth to students remained in flux. A stronger advocacy for manual training subjects among some educators only heightened this conflict and the challenges facing Eliot's committee. The Committee of Ten began its work in an effort to resolve these issues and to solidify and further the progress that had already been made.

# Chapter 5

# Charles W. Eliot and the Early Campaign for a National Educational System

## I

"I have read the report—every page." The document, of course, was the eagerly awaited report of the Committee of Ten on Secondary School Studies. James. M. Greenwood, the school superintendent in Kansas City, Missouri, read all 250 pages of the main report and the attached proceedings of the subject conferences, and he was not impressed. "I have looked over the immense array of names and the positions of these that have made these reports, and I agree with one statement made by Colonel [Francis] Parker [of the Cook County Normal School], that the report is a unique thing."[1] This bland response was the best thing he could say about a report that cost over four thousand dollars to produce, relied on the work of one hundred of the nation's best educational experts, and took over a year to finish.[2] Charles W. Eliot, who chaired the committee and whose educational views dominated much of the report, saw the document as the best hope for overcoming the gap between the secondary schools and higher education. It was an ambitious, sweeping plan to articulate the nation's schools and to establish an educational system that had so far eluded America's education reformers. Greenwood saw it as an attack on the public schools and on the very teachers who taught in them. He concluded,

> In behalf of the city teachers, who constitute the silent majority in this report, I will say that in New York, in Richmond, in San Francisco, in

Boston, in Cleveland, in Toronto, in St. Louis, in Chicago, or in any of the best schools of the country, the teachers are doing excellent work that these men say they are not doing; better than the gentlemen themselves can do, and yet they are prescribing for us, as they claim, and not for themselves. Let them take their own medicine first.[3]

It was not that the report engendered this criticism because it provided a radically altered vision for the future of American education or even that it introduced new concepts into the debate on the relationship between higher and secondary education. In many ways the report merely reflected much of the debate that had occurred over the previous decades. The committee had a loose mandate to examine the different subjects that were part of the secondary school curriculum and to hold conferences to address what to teach as part of these subjects and how best to teach them. But the report covered a wide range of issues. It tackled the difference between preparation for college and preparation for life. It laid out a plan to ensure that secondary school courses of study had an equivalence in preparation and mental discipline—a concept that measured a subject's worth according to its ability to demand rigorous mental development. The report also explored ways to unify the first years of the secondary school courses so that all students, regardless of their future directions, would study a common curriculum. Even though the committee hesitated to question the value of many of the subjects found in the typical secondary school curriculum, the members dealt implicitly with what it meant to be an educated person. They also waded into the preparation of teachers and called for more professional (i.e., college-educated) teachers for the high schools. For decades, educators had been discussing these ideas in countless education associations, meetings, and journals.

Eliot's committee was controversial, in part, because it was responding to the decentralized nature of education in the late nineteenth century. No central authority at the federal level had responsibility for America's schools. Although there was a degree of uniformity in education, especially following the spread of accreditation programs and the organization of some regional associations, educators bemoaned the pervasive differences that continued to hinder a complete uniformity across states and regions. Eliot pushed his committee into this void, hoping to establish a common set of principles and subject requirements according to which schools from coast to coast would operate. His faith in progress and in the overriding spirit of efficiency that followed vast industrial and commercial successes in a highly regulated and structured economy buttressed his belief that a strong, well-articulated, standardized system of schools was essential and crucial to the health of the nation. Educators wanted greater uniformity, to be sure, but they were not convinced that the report's recommendations were the right way to create that degree of standardization while also maintaining local traditions. Across the nation, then, the committee's recommendations rarely changed what secondary schools did, the content of their courses, or the focus of their work.

The committee was influential not because it dramatically transformed American schools or offered a decidedly radical vision for education. The committee built on earlier ideas, anticipated important initiatives that later gained prominence, and generated pronounced discussion. It failed to directly transform schools, except perhaps in a handful of cases, but it provided a rationale for national uniformity, advanced this agenda, and thus made it easier for later initiatives moving in the same direction to gain momentum. And, as historians have argued, it represented the first national attempt to carve out a central place for professional educators in setting the nation's education agenda. The Committee of Ten, then, was important because it represented a turning point in American education and in the campaign to create an articulated system of schools that looked remarkably similar from state to state.

This chapter starts with an overview of the committee's work and then considers the few cases in which it directly altered some secondary schools. The bulk of the chapter, however, focuses on the lasting role the committee played in the campaign to create a national system of schools. In this latter section, the chapter highlights three key ways in which the Committee of Ten advanced this system of education and affected future reform programs: it raised the issue of national uniformity in a forceful manner on the national stage and provided a way to ensure that the modern subjects gained equivalence with the classics; it highlighted the place of professional educators and experts in reforming education; and it underscored the influence that university-based teacher training programs were coming to have in American education.

II

Appointed in 1892 by the National Education Association (NEA), the Committee of Ten was the association's first significant foray into educational research and reform. Prior to this time, the NEA had confined itself to discussing pressing educational matters and offering weighty pronouncements.[4] The decision to move in this new direction and to appoint the Committee of Ten followed a report by James Baker that examined uniformity in school courses and in admission requirements, and a further conference of college and secondary school representatives on this topic chaired by Nicholas Murray Butler, a young professor at Columbia University. Principal of Denver High School and soon to become president of the University of Colorado, Baker called for such a conference to eliminate the diversity in subjects taught, teaching methods, and college admission standards prevalent throughout the country. "We have uniformity of government and institutions, of traditions and ideas." Likewise, "there is no reason why education of the same grade should not be substantially the same throughout the country," he claimed.[5] Baker was not going to be content with standardization in one state or even in one region; he wanted national uniformity. After all, students increasingly crossed state boundaries to attend college. "As soon as possible," he declared, "educators should make the question [of uniformity] a national one."[6] The Committee of Ten aspired to create such a standard of education that would apply from one coast to the other.

Eliot was quickly appointed as the chair of this committee and became its leading voice.[7] Along with William T. Harris, the U.S. commissioner of education, he was the most recognized member. Others on the committee included James B. Angell from the University of Michigan, and Baker, so instrumental in the committee's formation. Richard Jesse, president of the University of Missouri, and James M. Taylor, president of Vassar College, along with Henry King, a professor at Oberlin College, also represented higher education. The secondary school members included John Tetlow, headmaster of the Girls' High School and the Girls' Latin School in Boston, and James Mackenzie, headmaster of the Lawrenceville School in New Jersey. Oscar Robinson, principal of the high school in Albany, New York, was the only member to represent a public high school, although others on the committee had worked in the public schools earlier in their careers.[8]

As Eliot explained, the committee—supported with a $2,500 appropriation from the NEA—was "an attempt to bring together college and school men in consultation concerning the best methods of teaching each subject which enters largely into the programs of secondary schools, the proper limits of each subject, the best modes of testing attainment in those subjects, and the feasibility of attaining a tolerable uniformity of topic, method and standards throughout our wide country."[9] Eliot held out the hope that the committee would encourage greater cooperation among educators and solve the long-standing problem of the relationship between higher and secondary education, an issue that only had intensified as the public high schools came to dominate secondary education. In the absence of federal legislation, the Committee of Ten, he hoped, would be a way to establish some educational coherence on a national scale. Ambitious as ever, Eliot avowed that the report would "be an important contribution to the cause of education in the United States."[10]

The Committee of Ten set to work immediately. The members called for nine conferences of ten people each to meet and discuss key secondary school subjects. The members of these committees were experts in their fields, and the Committee of Ten selected them to balance higher and secondary education and to represent different geographical areas of the country. Forty-seven came from the colleges and universities, and forty-two represented the secondary schools (one was a government official).[11] Building on a notion of expertise reflected in Angell's inspection program and in William Rainey Harper's affiliation reforms at the University of Chicago, Eliot and the other members of the Committee of Ten ostensibly filled the conference committees with some of the best minds in each field. Few high school teachers found themselves appointed to the conference committees, however. The road to expertise went through the university and research or through the principal's office, not through the high school classroom and teaching experience. The secondary school representatives for the most part were principals, headmasters, and superintendents. These men were charged with creating courses of study for secondary school teachers, many of whom were female, and for students who also were often female. The committee reflected the administrative structure of education and not those who were in the classrooms.

In organizing these conferences, the Committee of Ten surveyed leading secondary schools in the country on the courses they offered. From this list, Eliot and his colleagues determined that the secondary schools taught nearly forty different subjects. The ten members of the overall committee used this information in deciding which subjects to focus on in the subject conferences, but Eliot never explained how the committee whittled forty subjects down to nine conferences. They likely chose the most prominent courses in many of the nation's secondary schools and colleges. After the report came out, Eliot declared that a handful of modern subjects, in addition to the classics, were the most popular courses among Harvard undergraduates. These courses, he felt, should therefore make up the curriculum of a good secondary school. The courses he identified were identical to the committee's nine subject conferences. There were separate conferences for Latin; Greek; English; other modern languages; mathematics; physics, astronomy, and chemistry; natural history or biology, including botany, zoology, and physiology; history, civil government, and political economy; and geography (physical geography, geology, and meteorology).[12] One member of the conferences questioned this limited number of subjects and asserted that the committee, in settling on them, "ran counter to the judgment of not a few." For instance, the principal of the New Bedford

High School in Massachusetts emphasized that "the friends of manual training are amazed that so important a subject should be disregarded."[13] From the beginning, however, Eliot had no intention of completely restructuring American education and arguing, for example, for manual training courses.

The members of these subject conferences met in December 1892 and, a few months later, presented detailed reports on a number of key issues. At the request of the Committee of Ten, each conference considered what needed to be taught in each subject and how best to teach the core requirements. They also considered how many years each subject should be taught. However, according to instructions from Eliot, they were not to support recommendations that went far beyond the "actual condition of American schools."[14] They had to work within the current context of education and not promote an idealistic curriculum impossible to implement in America's secondary schools. Eliot's committee also asked them to consider the extent to which the subjects should be considered as college admission requirements. Finally, addressing one of the most crucial questions, the committee wanted the conferences to wrestle with how best to teach students going to college and those not going to college. "Should the subject be treated differently for pupils who are going to college, for those who are going to a scientific school, and for those who, presumably, are going to neither," the committee asked.[15]

After the conferences considered such questions and reported to Eliot, Harvard's president read through and considered the reports and, with Tetlow's help, wrote the first draft of the committee's report. Eliot circulated the subject reports and the initial draft of the Committee of Ten report to the remaining members of the committee for their feedback. After revising this draft, he called for a full committee meeting to finalize the committee's recommendations, which met at Columbia University in November 1893.[16] *The New York Times* reported that the committee met in eight "secret sessions" of three hours each to investigate secondary education and make suggestions for its improvement.[17] The final report appeared at the end of the year, and it detailed a comprehensive set of recommendations for building strong schools that would ensure the health of the nation.

Most of the committee's recommendations came together in one table that outlined four potential courses of study for the secondary schools, with each course corresponding to one of the basic courses of study that students could pursue in college. The traditional Classical course focused heavily on Latin and mathematics but reduced the amount of Greek traditionally required. The Latin–Scientific course eliminated Greek entirely and focused more on the sciences. The course in Modern Languages replaced Latin and Greek with French and German and otherwise was identical to the Latin–Scientific course. The English course maintained one foreign language (either modern or classical) but focused on English, history, and science. The differences in the four courses were mainly with the type and amount of foreign languages required. They all expected students to have some exposure to English, at least one foreign language, mathematics, history, geography, and the sciences. They differed only in the amount required in each subject.[18]

By basing the courses on a common foundation, the committee sought to prevent students from having to make a choice about their future plans until their junior years. The first two years of each course were remarkably similar, and the Classical and Latin–Scientific courses were identical except that the Latin–Scientific replaced history in the Classical course with a science. The committee arranged these two courses in this way so that students could postpone until "the third year the grave choice between the Classical course and the Latin–Scientific," the report said. "This bifurcation should occur as late as possible, since the choice between these two roads often determines for life the youth's career." Eliot and the other members also affirmed that they felt it was impossible for students to make choices about their aptitudes and abilities until they had studied different subjects. "The youth who has never studied any but his native language cannot know his own capacity for linguistic acquisition; and the youth who has never made a chemical or physical experiment cannot know whether or not he has a taste for exact science," the report said. Another consideration guided the members of the committee. "Inasmuch as many boys and girls who begin the secondary school course do not stay in school more than two years, the Committee thought it important to select the studies of the first two years in such a way that linguistic, historical, mathematical, and scientific subjects should all be properly represented." The committee wanted to provide those students unable to continue in school with a rich program in the first two years that exposed them to the best of western civilization and prepared them for the demands of life.[19]

Relying on his own popularity in educational circles and the expertise of the committees he assembled, Eliot believed that he could use the committee and subject conference reports to lay the foundation for a unified course of study throughout the country. Moreover, since Eliot and the other members repeatedly emphasized that the four programs were suggestive of what secondary schools could do and were not prescriptive, the committee urged the colleges to accept any secondary school course of study for a corresponding college degree. Acceptable courses, however, had to adhere to the tenets laid down in the subject conference reports and had to represent a strong, unified, comprehensive curriculum. Eliot's committee was trying to provide a basis for national standards in each subject, while allowing schools and universities some flexibility in the exact arrangement of those subjects. Finally, addressing one of the great debates in educational circles, the Committee of Ten argued that any course of study—developed in adherence to the recommendations in the report—should prepare students for any college in the nation, as well as for the demands of life. Preparation for life and for college essentially could be the same.

In providing a foundation for a unified course of study that addressed the needs of students whether going to college or not, the Committee of Ten sought to resolve some of the most pressing educational issues and concerns of the day. For such an ambitious goal, it had very little direct effect on the nation's secondary schools. Eliot did have some success in altering Harvard's admission standards to meet the general tone of the Committee of Ten, and Harvard's faculty began to study the college's admission requirements in 1894, in light of "recent elaborate discussions of the objects and limits of secondary education and of the proper form and nature of the preparation for college." Harvard's professors, however, took their time in revising the requirements and did not report on a new scheme until 1897.[20] These new requirements did not adhere precisely to the recommendations in Eliot's report, but they did follow its general spirit and reduced the role of Greek and increased the place of the modern subjects. The faculty devised the new requirements, the dean of the faculty explained, as they "had been instructed to do, in conformity with the programmes of the Committee of Ten for preparatory courses of four years."[21] This change brought Harvard more into line with many secondary schools, especially the smaller ones, by allowing for a greater election in subjects. Smaller schools, constrained by their size and limited resources, now had the option of preparing students for Harvard without necessarily having to offer the full classical course of study. The Committee of Ten had supported similar reforms to make it easier for students from such schools to enroll in college.[22] Eliot also raised the requirements of the Lawrence Scientific School in line with the rigor and training needed for admission to Harvard. Creating such equivalence in preparation for different degrees had been a key goal of the committee's recommendations.[23]

The report also apparently had some effect on white secondary schools in the South, where some states—Georgia, Mississippi, Kentucky, and Missouri—claimed to develop their high school courses of study along the lines envisioned by the Committee of Ten. The South likely was willing to use it as a guide for improvement, since the region's secondary schools were in a poor state. For a region hoping to improve its schools, the committee and the conference reports provided useful suggestions. Whether at Harvard or in the South, however, the movement toward the modern subjects and closer alignment with the secondary schools preceded the Committee of Ten and its report. Eliot's committee often reflected the larger discussion and debate occurring in educational circles rather than inspiring schools to follow a new path. A fairly conservative document in that regard, it did not move far beyond its mandate and radically reenvision the state of American schooling. Instead, it reflected many of the trends already shaping secondary schools and colleges.[24]

Overall, few secondary schools or colleges adjusted their courses of study to match those outlined in the committee's report.[25] The University of Wisconsin, for example, declined, even after direct appeal by Eliot, to change its requirements to meet those advocated by the subject conferences and the committee. Instead, it decided to "devote its efforts to bringing the schools fully up to the courses of study now recommended by the State Superintendent before proposing further changes in the course," according to a faculty committee that considered Eliot's request. Both this committee and Charles Kendall Adams, the university's president, doubted whether the state's high schools could even meet the standards outlined in the committee's four courses. "The programmes suggested by the Committee of Ten demand more work in foreign languages than can be given in many of our schools," the faculty committee reported in 1894, "and that, as most Wisconsin schools are equipped now and must be equipped for sometime to come, our present requirements in physiology, botany, and physics are better adapted to the purposes of science teaching than is the greater variety of scientific subjects proposed by the Committee of Ten."[26] The state requirements also listed fewer subjects per term than the committee recommended, with the stipulation that these subjects should be studied for longer periods each week.[27] The state context trumped any movement toward national standards established by the Committee of Ten. Wisconsin's university was in no position to alter its requirements in line with the committee's recommendations if the state's high schools were pursuing a different path.

The report garnered significant press and discussion, but few superintendents or boards of education felt compelled or able to reform their schools in line with its suggestions.[28] Edwin Dexter, director of the School of Education at the University of Illinois, concluded in 1906 that "the report of the Committee of Ten seems not to have influenced directly to a marked degree the curriculum of public high schools."[29] In New England, John Tetlow tried to build support for the report. He sponsored a resolution through the New England Association of Colleges and Preparatory Schools that recommended that the secondary schools adopt the courses outlined by the committee as the basis of their work. This resolution ran up against adamant opposition from Greek professors, led by the Harvard

Greek department, and classical supporters who decried the reduced role that Greek claimed in the model programs. Even with the support of Eliot who rallied to Tetlow's defense, he failed to convince New England to embrace the Committee of Ten's recommendations. Faced with such opposition, Tetlow offered a weakened resolution that, although approved, was largely symbolic.[30]

Perhaps the most significant indication of the limited effect of the Committee of Ten came in 1895 when the NEA appointed another committee of experts to determine how to articulate secondary and higher education and to lay a foundation for the standardization of education. The purpose of this new committee, echoing the work of the earlier committee, was to harmonize "the relations between the secondary schools and the colleges, to the end that the former may do their legitimate work, as the schools of the people, and at the same time furnish an adequate preparation to their pupils for more advanced study in the academic colleges and technical schools of the country."[31] The Committee of Ten had not succeeded in resolving the debate between preparation for life and college or in creating the foundation for a stronger relationship between the two educational levels. Ironically, however, the committee had succeeded in one crucial aspect. Although its effect on the schools was proving to be limited, it had crafted the model for committees that followed. When the Committee on College Entrance Requirements began its work in 1895, it hewed closely to the structure first developed by Eliot's committee. Its membership included educators from higher and secondary education, although a secondary school superintendent—Augustus F. Nightingale of Chicago—chaired the committee. It also worked with committees of experts to study subjects and recommend how best to teach them. In an important departure, however, it did not establish separate subject committees but worked with professional associations—including the American Historical Association, the Modern Language Association, and the American Mathematical Association—to develop standards for their disciplines.[32] As this new committee evolved, the members of the Committee of Ten had to be disappointed in the lack of progress made in reforming the nation's schools following the report's publication.

The Committee of Ten addressed some of the thorniest issues of the day and packaged them into a broad, comprehensive program of reform. Although the committee did not originate many new ideas, it did combine a number of ideas into one program, and it proposed that package as a national agenda for educational change. Eliot and his committee, eager to solve many of the problems plaguing schools and education, produced a document that attempted to formalize many of the changes already occurring in various states and regions. It was an ambitious document that envisioned a strengthened system of education leading up from the second-ary schools and into the colleges and universities. It also attempted to find solutions to contentious issues that were still being discussed and that engendered significant debate. G. Stanley Hall, who initially said positive things about the report, put it succinctly in 1894 at the NEA's annual meeting. "It interferes with so much, however, that I am not surprised that it should meet with opposition."[33] That the report tried to bring together a number of vital issues and to provide a sweeping solution meant that it ignored opposition and circumvented debate among people who were not yet ready to agree on the direction of American education. Not surprisingly, it garnered little support across the country.

Even though based on the best advice of one hundred experts, the report had little hope of radically altering the nation's schools. Most educators in the country were not in agreement on what shape the schools should take. Eliot was ready for resolution, but most of the nation's teachers, superin-tendents, professors, and presidents continued to disagree on the best path for the schools. In this context, the best that could be expected was what some of the committee's supporters hoped for—discussion. It certainly spurred discussion around issues central to articulation. Angell claimed that this debate was what the committee really wanted from the docu-ment. "I think I do not misrepresent the Committee when I say that they did not flatter themselves that their work would be accepted without criti-cism, but that they did hope that it would awaken intelligent and earnest discussion through the country," Angell wrote in 1894. The committee members hoped that discussion by teachers, "who had carefully studied the very valuable reports of the special conferences, would arouse fresh interest in the problems of secondary education, and lead ultimately to the great improvement of our high schools and academies." Based on what he saw around him, Angell believed that, if nothing else, this expectation had been "well founded."[34] Burke Hinsdale, Angell's colleague in Michigan, agreed that the report's "most valuable service will prove to be its lifting the whole subject of secondary education up into the clear light of public knowledge."[35]

They were not wrong. The Committee of Ten became the subject of education meetings from state to state, and a number of state and national journals launched discussions on the merits of the committee's report. Prior to the committee's efforts, articulation had been primarily a state and regional issue. Eliot's report brought articulation onto the national stage, and three of its points proved particularly crucial to the campaign to articulate higher and secondary education. First, it promoted national uniformity in the secondary schools and in college admission standards, and at the time raised these subjects to their highest level of attention.

Reaching this goal of uniformity demanded that higher and secondary education reconcile the dual demands placed on the high schools. It also required that Eliot's committee find a way to ensure that the modern subjects equaled the classics in academic rigor and preparation. Second, it highlighted the ongoing role of experts, primarily university-trained experts, in the work of the secondary schools, and it gave them a national stage from which to promote their views. Finally, it underscored the growing role of university-based teacher training programs in American education.

# III

Most attempts to articulate the two educational levels prior to the Committee of Ten had occurred at the state level—through the inspection program in Michigan, for instance—or at the regional level where midwestern universities adopted the accreditation system. The New England Association represented an alternative movement toward regional unity, but the Committee of Ten was the first document to argue for national unity and to provide a way to bring that about. James Mackenzie, headmaster of the Lawrenceville School and a member of the Committee of Ten, spoke for many in 1894 when he complained about the lack of uniformity and the evils of diverse and shifting college entrance standards. "The utter chaos into which college entrance requirements have fallen—the revealed idiosyncrasies of the college faculties often ruthlessly enforced upon long-suffering, protesting schoolmasters,—is a railing reflection upon the intelligence, good sense and fair play of the American people," he proclaimed in terms that echoed what educators had been saying for decades.[36] He saw the Committee of Ten as a way to reduce the suffering of his colleagues in the secondary schools.

As Americans looked around them, they saw the benefits—transportation and communication improvements, as well as new businesses—of an industrial system seemingly operating according to norms of efficiency and standardization. Education reformers may not have spoken of the schools in stark, industrial terms, but across the nation they worked to create a more efficient, regimented structure. Each grade, they claimed, should have its work to do and should do that work fully and efficiently, whether in Los Angeles, Ann Arbor, or Boston. Teachers should take their students and send them up the ladder to the next grade, until throughout the nation students as a matter of course poured out of the schools at graduation in the spring and entered college or the business of life.

The Committee of Ten attempted to provide the basis on which this nationalized, structured system could further evolve. Table IV of the committee's report, which contained the four model courses of study, perhaps inevitably drew the most attention. Embodied in these four courses were all of the recommendations of the committee condensed into an easily digested format. Educators quickly concentrated on them as the experts' principal recommendations, much to the dismay of Eliot who saw the model programs only as examples of what schools could do. Instead, Eliot hoped that schools would build their own programs based on the recommendations of the report. The conferences and the final committee report provided detailed information on the proper subjects for the secondary schools and on the topics that should be included in each subject, the best ways to teach a subject, and the appropriate time to be allotted to each subject per week. Eliot wanted the schools to choose freely from these subjects in designing their own courses of study. As long as principals and superintendents throughout the country adhered to the recommendations in the report, Eliot believed, "there might arise in this way uniformity of method all over the United States," even if "the programs made for particular schools, or separate communities, should vary." These administrators could pick and choose the subjects to be included in the curriculum, but, as Eliot explained, "the boy who studied algebra, for example, in San Francisco, would study it as long, would cover as many subjects in algebra, and would be instructed in about the same way, as would a boy in Boston who studied the same subject." Uniformity did not mean that the courses of study in Boston mimicked those in Atlanta, Minneapolis, or Denver. Cognizant that local expectations often dictated what high schools could do, Eliot asserted that it would be "impossible to hope for one single uniform program in secondary schools." Nor did he think such uniformity was even desirable.[37] Rather, uniformity meant that students leaving Dallas and enrolling in another school in Bay City, Michigan, or in Nashville should find many of their classes being taught in a familiar way. Eliot hoped that the committee's recommendations would move the nation's schools in this direction.

Regardless of the exact nature of the course of study that students throughout the country pursued, if it followed the committee's recommendations, it would be a credible secondary school course, according to Eliot's group. But the report went beyond this recommendation and proposed that any such course should also be sufficient for college admission. Encouraging more students to enter college clearly motivated Eliot, as he made apparent in a letter in 1895 to Seth Low, president of Columbia University. "As you may have read between the lines" of the report, "I am very much interested in promoting a great widening of the requirements for admission to our colleges."[38] A uniform course of study that included more than the classical subjects and that rested on the recommendations in the report, he trusted, would succeed in broadening the road leading up to the colleges. "In order that any successful graduate of a good secondary school should be free to present himself at the gates of the college or scientific school of his choice," the report proclaimed, "it is necessary that the colleges and scientific schools of the country should accept for admission to appropriate courses of their instruction the attainments of any youth who has passed creditably through a good secondary school course, no matter to what group of subjects he may have mainly devoted himself in the secondary school."[39] In this way, the Committee of Ten sought to provide the basis for ridding the secondary schools of the challenge of fitting students for different colleges.

For the elusive goal of eliminating diversity in entrance standards to become reality, however, secondary and higher education both had to be willing to follow the recommendations contained in Eliot's report. The secondary schools needed to create at least one rigorous course of study in line with the report's suggestions, although the larger schools might choose to offer separate programs in preparation for the various classical, scientific, and English degrees. The onus then was on the colleges to accept any such course of study. Under the plan envisioned by the committee, the secondary schools no longer needed to prepare students differently for any number of colleges. Eliot's solution, of course, relied on the colleges giving up some control over their admission standards and accepting a unified set of requirements for each subject. The sought-after uniformity did not imply that colleges had to agree on the subjects required for admission. They only had to agree that, where they accepted the same subjects, they would set identical standards. Eliot advocated for such uniformity long before he became chair of the Committee of Ten, but this national platform gave him the stage from which to further push his goals.[40]

Promoting a unified basis for admission to college motivated Eliot and his committee. Finding the best way to prepare one set of students for college and another for the demands of life further roused them to action. Educators had long debated in journals and meetings how best to prepare students for different paths. Should the secondary schools offer programs designed to send students out into life with a strong foundation for citizenship, while also fitting students for college, they wondered? Would an identical curriculum for both sets of students meet the needs of the colleges and of life?

When the NEA formed the Committee of Ten in the early 1890s, educators had not yet definitively answered these questions, and the members of the committee understood the problematic position of the secondary schools. The "main function" of these secondary schools, the report declared, "is to prepare for the duties of life that small proportion of all the children in the country—a proportion small in number, but very important to the welfare of the nation—who show themselves able to profit by an education prolonged to the eighteenth year, and whose parents are able to support them while they remain so long at school." The committee members concurred then that preparation for college was to be "the incidental, and not the principal object" of the secondary schools. Still, they maintained that it was "obviously desirable that the colleges and scientific schools should be accessible to all boys or girls who have completed creditably the secondary school course."[41]

In other words, the committee wanted the schools to fulfill both purposes. In an attempt to figure out how the schools could reach both ends, Eliot and his committee asked the subject conferences to determine whether students should have a different education depending on their future paths. Without exception, these conferences answered that all students should receive the same education in those courses they had in common. It was to make no difference, they claimed, whether one student was going to college and one was not. As long as students took the same course—algebra, for instance—they should receive the same education and be treated in the same way for the length of that course. The members of the conferences, the report trumpeted, "unanimously declare that every subject which is taught at all in a secondary school should be taught in the same way and to the same extent to every pupil so long as he pursues it, no matter what the probable destination of the pupil may be, or at what point his education is to cease."[42]

The implication of this uniformity in preparation was that the courses laid down by the committee were suitable to fulfill not just college entrance standards but also life's many demands. What was good enough for the students leaving high school for life's pursuits should also be good enough for those going on to college, the committee seemed to be saying. As Eliot explained in 1894, "The Committee clearly desired to establish a closer connection between secondary schools and colleges, and therefore made a general recommendation to the effect that the satisfactory completion of any good four years' course of study in a secondary school should admit to corresponding courses in colleges and scientific schools." While hoping to strengthen the connection between higher and secondary education and encourage more students to enroll in college, the committee members also made it clear that, "in their judgment, a secondary school programme intended for national use must be made primarily for those children whose education is not to be pursued beyond the secondary school."[43] Preparation for college rested on the same foundation as preparation for life, in the committee's formulation, and the needs of the greater number not going beyond high school were to dictate the course of study for the few progressing beyond secondary school. Crucially, however, the members disagreed on whether preparation for life should be the basis of preparation for college or whether preparation for college formed the foundation of preparation for life. Eliot seemed to imply that preparation for life was the guiding factor, but his colleague on the committee, Henry King, argued that "instruction that fits for college almost equally fits for life."[44] This distinction was more than a rhetorical point, since it went to the crux of the matter. In attempting to resolve this debate, the committee was never entirely clear whether the needs of college or of life were the motivating factors in the development of the curriculum.

A unified course that met the requirements for college and for life and that was available to all students throughout the country would ensure that the road leading to higher education was as wide as possible without destroying the path to life. Students and their parents then could make the choice to enter college late in high school and not fear that they would be unable to meet the entrance requirements. Eliot's committee answered the pressing questions of how best to prepare students for life and college by arguing that any course of study developed in accordance with the committee's recommendations would meet the needs of all students, regardless of what the students did once they left the secondary schools. The committee affirmed the four-year course of study as crucial for students, whether those students went to college or into life's pursuits.

This resolution was not a wholly satisfactory one. H. S. Tarbell cautiously approached the notion that preparation for college and life were the same. Appreciating that such an idea made it easier for the secondary schools, he doubted that it provided the best education for students. "The purpose for which a study is pursued," he explained in 1894, "must influence the selection of the topics treated under that subject." For example, he continued, "Latin in the class preparing for college requires one treatment; but quite a different treatment in the class using it merely as an aid in knowing the sources of our literary vocabulary."[45] He did not know how a student preparing for college and another who had no ambitions for advanced study could approach the same course in an identical way. A student's purpose or goal for a subject determined, he thought, how it should be taught and studied. Moreover, as others complained, there was little room in the secondary school curriculum for manual subjects, stenography, bookkeeping, art, drawing, and music. The committee did allow that some commercial courses could be offered as alternatives, but manual training had no significant place in its report.

Calvin Woodward, a nationally prominent educator and head of the Manual Training High School in St. Louis, wondered why manual training was ignored in Eliot's report. Secondary schools were growing quickly and drawing more students to them, including those attracted to the manual subjects as preparation for life rather than to the modern and classical subjects contained in the committee's reports. These students, Woodward argued, had no advocate on the Committee of Ten. Mackenzie, who had been on the committee, agreed with Woodward that these subjects needed a place, and he insisted that he had raised these same objections during the committee's deliberations but had been "voted down."[46] Robinson similarly pushed for the manual subjects, art, music, and drawing.[47] Both men, representing secondary schools, raised concerns that ultimately were not reflected in the final report. Preparation for life and for college could be identical as long as this preparation included "academic" subjects—both modern and classical—and excluded such "practical" classes as commercial arithmetic and drawing.

For the modern and classical subjects to be acceptable as preparation for life and for any college, they needed to be based on a set of common standards and criteria. The committee recognized that such uniform standards did not exist and that many of the nation's secondary schools were in a rudimentary state. Although the number of secondary schools had grown considerably over the last decades of the nineteenth century and numbered 4,500 by 1890, these schools were not all of high quality. "The pupil may now go through a secondary school course," the committee recognized, "of a very feeble and scrappy nature—studying a little of many subjects and not much of any one, getting, perhaps, a little information in a variety of fields, but nothing which can be called a thorough training."[48] The Committee of Ten knew that it could not ask the University of California, for example, to accept a Missouri student's high school preparation unless it could guarantee that the training was of a high standard and matched California's expectations. It had to provide a way to ensure that what was taught in a high school in Michigan closely resembled what was taught in a high school in New York. The only way to ensure this equivalence or uniformity in preparation was to establish a common set of requirements for all of the subjects that made up the secondary school curriculum.

The Committee of Ten believed that it had created such requirements in the subject reports that ninety experts had helped to write. These reports detailed what topics each subject should include and provided advice for teachers on how best to teach them. Eliot wanted the nation's teachers to have access to documents that clearly outlined what they should do in their classrooms. He often claimed that the subject reports were the strongest parts of the overall Committee of Ten report, and he and others encouraged teachers to refer to them often. "These Conference reports," Eliot stated, "contain a great number of recommendations for the improvement of teaching, not only in the secondary schools, but also in the elementary."[49]

The subject conferences gave teachers throughout the country an idea of what to teach. The "time-allotment" or the periods per week given to a subject—with forty-five minutes as the standard length—was the committee's way to ensure that the amount of time spent per week on each subject in one school closely resembled the amount of time spent in the rest of the nation's secondary schools. The committee wanted to avoid the spectacle of one school in Massachusetts offering English for three periods a week in the first year, while the rest of the state's secondary schools offered English for four periods. The committee needed a consistent foundation for the content of subjects and it also needed a basis for quantifying that the amount of history taught in Wisconsin equaled that taught in Michigan. In developing the model courses of study, the committee members settled on four as the appropriate number of periods per week to be spent studying most subjects. The first year of a language course got five periods and some subjects got only two or three periods per week, but four periods per week was the average for each subject.[50] Combining the subject conferences as the basis for the quality or content of the courses with time-allotment as a way to measure or quantify the time spent in the courses provided a basis for equalizing teaching across the country.

But if colleges were to accept the modern subjects in lieu of the classical courses, they needed to know that a student's preparation in, say, English had been as rigorous as it would have been had the student taken Latin. To argue that any two subjects, especially modern and classical, had an equivalent worth, Eliot and the committee turned to a long-standing faith in mental discipline. By the 1890s and the time of the committee's report, mental discipline was a somewhat archaic concept, but the committee's authors championed mental discipline as the basis of a sound course of study. For a subject to embody the best of mental discipline and thus earn a place in the curriculum, it had to help students make accurate observations, record and order those observations, make judgments about them, and then draw appropriate inferences from their observations, classification, and analysis. The content of a subject was secondary to the reasoning powers it developed. For many, the classical subjects of Latin, Greek, and mathematics provided the strongest mental discipline.[51]

Eliot wanted to bestow on the modern subjects a mental discipline equivalent to that of the classics. As he told Columbia's president, he wanted to encourage more rigorous requirements in history, science, and other subjects so that the colleges would readily accept these subjects for admission on a par with the classical courses.[52] The recommendations of the subject conferences, the concept of time-allotment, and the idea of mental discipline gave him a way to reach that goal. As the report said, "If every subject is to provide a substantial mental training, it must have a time-allotment sufficient to produce that fruit." All of the subjects from which a student is to choose, it continued, "should be approximately equivalent to each other in seriousness, dignity, and efficacy." "The Conferences," the committee reported, "have abundantly shown how every subject which they recommend can be made a serious subject of instruction, well fitted to train the pupil's powers of observation, expression, and reasoning."[53] As a result, two or more subjects taught for the required hours a week, in line with the dictates of the conference recommendations, were equivalent in their ability to train the mind. Only in this way could Eliot and the committee argue that English should be as acceptable in meeting college admission standards as Latin. After all, a set number of hours a week for one year in both subjects provided equal discipline and training—of course, only as long as both subjects followed the recommendations of the subject conferences. A few years later, the North Central Association, a regional organization of states in the Midwest and West, and the Carnegie Foundation for the Advancement of Teaching built on these ideas and established the "unit" as a measure of the time spent studying a subject and the content of that subject.

# IV

Through its trust in educational experts, the Committee of Ten highlighted the growing professionalization of teaching and the reliance on expertise that was occurring in education and throughout much of the country. The committee's report was a showcase of expertise. Scholars and experts representing more than a dozen subjects came together in conferences to formulate the content of those subjects and the best way to teach them. They represented professors from the academic disciplines, the new breed of education professors, university presidents, and secondary school administrators. Education, of course, was not alone in its reliance on well-trained experts in the late nineteenth century. Throughout a society nervously wondering how to deal with the changes wrought by the interrelated forces of urbanization, industrialization, and immigration, Americans turned to experts. The Committee of Ten was an opportunity for educational experts to shine on a national stage and to create the framework for dramatic reform and for the emergence of a unified, fully functioning system of education.

Embedded in this call for expertise and national uniformity was a ringing disdain for the lay school boards that imposed provincial control over the schools. Ward bosses and local politicians often controlled these boards and, according to professional reformers, hired incompetent teachers, wasted money, accepted bribes, and generally corrupted the schools. Reform-minded educators hoped to replace these current boards with reformers dedicated to sound educational principles.[54] Even in cases where party bosses did not have control, the development of strong secondary schools could still be undermined by local expectations. "The secondary schools themselves, not always conducted in a wise or generous spirit," Nicholas Murray Butler wrote in 1894, "have too often sacrificed the necessities of sound training to the local demand for an ambitious programme containing twoscore or more of school subjects." Or, he claimed, "they have erred on the other side, and in their devotion to a past ideal excluded from the curriculum whole fields of knowledge that have grown up within a century." Columbia's young professor neglected to mention that the colleges had long "erred" on the side of the classical ideal. Clearly, for him and others—including Eliot—however, the schools needed the leadership of an expert body of educators trained in universities by the nation's top minds and attuned to the best that education could be. The Committee of Ten represented a national attempt to make the expertise of professional educators the basis for what occurred in the secondary schools. Indeed, Butler judged that the members of the conferences were "so admirable" that it was difficult for him "to speak of them without enthusiasm." Many of the ninety members, he claimed, stood "in the foremost rank of American scholarship" and all had valuable educational experience that would serve them well.[55]

Historians have addressed the growing professionalization of teaching in the 1890s, and some have argued that the Committee of Ten was the beginning of an attempt to assert the authority of professional educators over education and to take control of the schools out of the hands of lay boards. Although the Committee of Ten certainly represented the strengthening power of professional educators, it was not the beginning of this process.[56] Rather, this trend started earlier and can be seen in the Michigan inspection and accreditation program. Well-trained experts in their fields visited schools and made recommendations for improvement. Some even worked with reform-minded principals and superintendents to move headstrong boards toward change. At the time of the Committee of Ten report, Harper at the University of Chicago had expanded on the role of university professors and expertise, and built a model where the university worked to gain significant control over the educational mission of secondary schools. The Committee of Ten reflected on a national level the strengthening role of university professors and reform-minded administrators in controlling the schools—a movement that had been occurring at state and regional levels. The committee, however, did represent a departure from the earlier models in an important way: it used professional educators to develop detailed content for subjects and promoted the subject reports as invaluable aids in classroom teaching. Michigan's inspectors—and professors from other universities as the program spread—had sought to influence classroom teaching and they certainly suggested appropriate textbooks to use, but they refrained from offering detailed reports on subject content.

As part of the Committee of Ten report, presidents, professors, and secondary school administrators (with only a handful of teachers) came together to offer advice on reforming education throughout the country. These professional educators did not always agree on the future direction of education, but they were united in believing that they, rather than lay school boards, should have control of schools.[57] Eliot and his committee understood that lay boards and local needs traditionally dictated a school's curriculum, and Eliot thought that such flexibility was important. The Committee of Ten, nonetheless, wanted to reduce local control and place it within a larger framework of national uniformity. Flexibility could occur around a core group of subjects, the committee's members conceded, but their work was an attempt to reduce the influence of these lay boards by promoting a national idea of what secondary schools should be. They provided some flexibility to allow local schools to offer German instead of Greek or add a class in commercial arithmetic. The experts on the Committee of Ten assumed lay boards would tinker somewhat with the course of study but opposed significant revisions to their work.

The Committee on College Entrance Requirements similarly relied on experts to determine the content of subjects and how best to teach them. These expert educators and the members of the professional associations that developed standards in the academic subjects followed the lead of the Committee of Ten in challenging lay boards and local expectations for the public schools. One of the committee's central recommendations focused on the need for a national unit for measuring the amount of work done in a subject. The Committee of Ten also had sought to create such a standardized basis for measuring and evaluating subjects throughout the country. Nightingale's committee promoted the unit as a basis for "determining the amount or quantity of such subjects or studies as shall be required by the college or the school."[58] The committee, through the professional associations, also stipulated the content of each subject, including the quality of the work done and the methods of teaching them. The unit, then, represented the quantity and quality of the academic subjects and the best way to teach them. It meant a definite amount and value of work done in each subject.[59] "The aim of the Committee on College Entrance Requirements," the report stated, "is to set forth such a series of interchangeable units of substantially the same value as will meet with acceptance everywhere." Although local conditions might dictate different groupings of subjects, each subject studied for a certain number of units would carry the same weight in Chicago as in Denver. "That is to say," the committee said, "one unit of history taught in one place should equal one unit of history taught in another place, even tho [sic] the subject-matter of instruction varies."[60] The committee hoped that colleges would state their entrance requirements in terms of units and that the secondary schools would establish programs in line with unit requirements. Nightingale's committee, through the work of the professional associations, essentially followed the Committee of Ten's recommendations and provided a basis on which higher and secondary education could begin to establish quality programs in each subject. It outlined the unit basis for all subjects, or the "quantity, quality, and method of the work in any given subject of instruction."[61] The Carnegie Foundation for the Advancement of Teaching later built on this movement toward a national, standardized means of measuring school attainment. Direct change failed to follow the Committee of Ten's work, but later experts and committees built on its ideas.

Critics repeatedly claimed that these experts had little authority for evaluating the secondary schools and creating a course of study for them. For some secondary school educators, this reliance on experts bordered on college domination. "The tendency of the day," according to Greenwood, the superintendent of schools in Kansas City, Missouri, was "to put the schools, at least the plans therefor, in the hands of specialists who are putting upon them burdens that they cannot bear."[62] Clearly not all school administrators supported the work of the expert professional educators who were seeking to control America's schools. More so, perhaps, than other urban superintendents, Greenwood had faith in his teachers and their ability to conduct their classrooms and teach their students in a strong, pedagogically sound manner.

Caskie Harrison, of the Brooklyn Latin School, understood Greenwood's point. Of the ninety individuals appointed to the conference committees, forty-two came from the secondary schools. This distribution troubled Harrison. "The easy assumption of propriety in this distribution is enough in itself to show that, in the eyes of our Colossi, we must be underlings even in the measure of our authorized aspiration; but our subserviency is far greater," Harrison contended. The main problem for Harrison was that of these forty-two school men, nearly two-thirds were principals. Secondary school teachers had little if any place in the work of the committee and the conferences. "It is fair to assume," Harrison concluded, "that not more than half of the total number of school men have been actual teachers so recently and so largely, or have otherwise lived in such close and unbroken relations with the real work of teaching, as to be teachers in any true sense." Instead of leaving the work of the conferences and the overall committee to so-called professional experts, it should have given secondary school teachers a prominent role in formulating the committee's conclusions, he argued. "Teaching is a practical matter, with conditions of system and sequence and completeness and repetition and illumination, from one or all of which most college-teachers," Harrison claimed, "consider themselves exempted." College teachers should have given way to the expertise of secondary school teachers.[63]

M. A. Whitney, the superintendent of schools in Ypsilanti, Michigan, concurred that expertise had to be redefined to include a broader array of people. "Our high school courses of study are dictated largely at present by specialists in the various branches," he declared, "and I am not ready to admit that they are always the broadest minded men." Although Whitney thought that the work of the subject conferences had value, he stressed that "nearly all who sat in those conferences clearly demonstrated that they were not the proper persons to construct our school program." He wanted a course of study modeled on strong pedagogical principles but not one dictated by university professors. "We are not ready to accept a ready made article even from our professors of pedagogy, any more than we are ready to do it from the specialists in our universities." For Whitney, the best course of study embodied "the best thought of the best educators, and they do not all occupy chairs of pedagogy by any means." He envisioned a more expansive basis for constructing the curriculum. His process embraced the expertise of a wide range of educators. "Let our pedagogical friends furnish the theory," he suggested, "and our best superintendents and teachers, furnish the practice, and our specialists furnish something of method, and I think together we might construct a very respectable course of study, which all will be pleased to adopt." For him, Harrison, and others, expertise had to include the contributions of those who spent their days teaching in the schools.[64]

Eliot, not surprisingly, disagreed with these attacks on the expertise of the one hundred educators and, specifically, on the college men who made up the overall committee and the conferences. All levels of education—whether kindergarten, high school, or college—were unified by a common set of principles and by a focus on training the individual "to see straight and clear; to compare and infer; to make an accurate record; to remember; to express our thought with precision; and to hold fast lofty ideals." From beginning to end, Eliot contended, education was "a continuous process of one nature." He argued that teachers from the lowest elementary grades to the graduate schools were engaged in the same line of work and should focus on the crucial goals of educating students to think and communicate effectively.

> Shall we not agree that there is something unphilosophical in the attempt to prejudice teachers of whatever grade against the recommendations of the Committee of Ten and of the Conferences that committee organized, on the grounds that a small majority of the persons concerned in making them were connected with colleges and that the opinions of college or university officers about school matters are of little value?[65]

For Eliot, education essentially adhered to the same principles regardless of the level. As such, college men and their expertise had an obligation to be involved in what occurred in the lower grades, and criticisms of such involvement were without merit. This expertise, rooted in their work at the highest levels of the educational ladder, was absolutely crucial to education reform, Eliot claimed.

# V

The Committee of Ten embraced such expertise in its campaign to define the role of secondary education and the specific subjects and content that should constitute the secondary school curriculum. The next step was to extend this expertise to teacher education and training. The whole scope of the committee's efforts to create unified secondary school courses that met the needs of students going to college (regardless of whether they enrolled in a classical, scientific, or modern course of study) and those going out into life rested on significant improvements in the training of teachers in line with the work outlined by the experts on the subject conferences. "Every reader of this report and of the reports of the nine Conferences will be satisfied that to carry out the improvements proposed more highly trained teachers will be needed than are now ordinarily to be found for the service of the elementary and secondary schools." Eliot and his colleagues, therefore, called on the colleges to take a more active role in preparing the nation's future teachers. By having some control over the training of teachers, both in disciplinary content and in pedagogical skills, higher education gained a powerful tool in exerting an influence over what teachers taught in the nation's secondary school classrooms and how they did their job.[66]

Having responsibility for training administrators might give the universities even more power in shaping secondary schools. The colleges, the report trumpeted, "ought to take pains to fit men well for the duties of a school superintendent. They already train a considerable number of the best principals of high schools and academies; but this is not sufficient."[67] Colleges and universities needed to take a more active role in molding the men who controlled and ran the nation's secondary schools. If they succeeded in this goal, they might eliminate the opposition that some superintendents and administrators—such as Greenwood—had to the committee's recommendations. These well-trained administrators would then join a growing group of professional educators dedicated to sweeping away vestiges of lay control and developing rigorous courses of study grounded in the recommendations of the Committee of Ten.

In its call for colleges to take a greater role in training teachers and administrators, the Committee of Ten, as was the case with many of its recommendations, highlighted what some colleges and universities were already doing. Many of the state universities had recently taken important steps in their efforts to prepare teachers and administrators for the public schools. In 1879, the University of Michigan established the nation's first "Chair of the Art and Science of Teaching," which Angell believed was necessary "to aid in preparing our graduates to teach in our schools or to superintend schools."[68] Students, he argued, needed "to be familiar with the principles which should govern the administration of such schools, with the philosophy of teaching, and with the history of education."[69] The added benefit was that a university-based education program would be able to imbue future educators with an appreciation for higher education and knowledge and with a desire to encourage students to attend the university.[70] In 1891, Eliot encouraged Harvard to offer courses in teacher education, and he hired a new professor to take charge of this work. Eliot was not entirely sure what he wanted this new professor to do, although he vaguely defined a number of duties: delivering lectures on the art and history of teaching; visiting schools and making recommendations for improvement; and conducting summer schools for teachers. Like Michigan, Harvard was moving toward formal teacher preparation, although it lagged behind Michigan by over a decade.[71]

The creation of a chair in the art and science of teaching at Michigan and in universities that followed gave higher education a direct role in the preparation of teachers. Summer schools for teachers additionally provided the universities an opportunity to mold and shape the teachers who prepared students for advanced study in higher education. These summer courses met the needs of practicing teachers, many of whom pushed universities to develop such programs. The University of Wisconsin opened its summer school for teachers in 1889, after the state legislature appropriated one thousand dollars for that purpose. By 1894, the summer school enrolled over 120 students, although attendance declined in the following years.[72] Prior to opening the summer school, the university sent its professor of pedagogy to deliver lectures throughout the state. These lectures on educational topics, the president claimed, contributed to the heightened "interest in the University shown by the teachers and people of the state, the large number of schools that have sought a place on the accredited list and the improved preparation of students."[73] Michigan followed Wisconsin's efforts, and opened a summer school in 1894.[74] As early as 1874, Harvard also had turned to summer courses, although not on a systematic basis, as a way to train teachers. Concerned that teachers in the secondary schools lacked the necessary skills to teach scientific courses by "rational methods," Harvard offered summer courses in chemistry and botany to help these teachers embrace the teaching of science through laboratories and experiments.[75]

Creating a new chair or opening summer schools eventually led to the development of full departments of pedagogy and schools of education. The University of Wisconsin took a significant step in teacher education when it launched a School of Education in the mid-1890s. At the request of the Wisconsin State Teachers' Association, the university opened this school to provide advanced courses in the science and art of teaching. The early success of the school, the president pronounced, "seems already to have abundantly justified its organization." A staff of six professors taught courses that had a total enrollment of nearly three hundred students during the school's first semester. Some of these students entered the new school after having graduated from the state's Normal School, but the school also worked with students from other departments of the university who hoped to gain some pedagogical skills before applying for positions in the state's high schools.[76] Six years after opening its doors in 1892, the University of Chicago similarly founded a college for teachers in the area's schools. Harper, the university's ambitious president, believed that rigorous training in the academic disciplines, with some focus in pedagogy and philosophy of teaching, best ensured that secondary school teachers gained the skills necessary to be effective educators.[77] By the early 1900s, university officials claimed that almost all of the schools accredited by the university had at least one teacher who had attended the university's college for teachers.[78]

Wisconsin, Harvard, Chicago, and Michigan, along with other universities, began to create a science of education in the late nineteenth century through their focus on summer schools, their employment of education professors, and their development of education schools. Butler, at Columbia, believed that these initiatives were crucial. He reasoned that universities had "a most imperative duty to the teaching profession,—the careful and systematic exposition of education considered as a science." For Butler, only the university could engage in such critical study of schools and teaching. "The university, and the university alone is equipped by tradition, by scholarship, by resources, and by opportunity to give to the subject of education that profound and accurate treatment that has characterized its study of the sciences, both moral and physical, during the past five hundred years," he asserted in 1890. As a result, the university needed to "construct a science of education from which the principles of the art of teaching will be readily derived." Butler placed his faith in such a science of education to reform the whole structure of American education and to create a harmony among all of its disparate parts.[79] Private and public universities across the nation had taken important steps to accomplish Butler's dreams, and, as far as training new teachers and administrators was concerned, to implement a crucial recommendation in the report of the Committee of Ten. They hired professors to train new teachers and they studied education. These universities and their professors worked with high schools and teachers on a regular basis, but the direction flowed from the universities into the supposedly empty vessels that were the teachers.

The Committee of Ten, in its call on a national stage for better training for teachers and administrators, reflected the growing role that colleges and universities were taking in preparing students for positions in secondary education. Gradually this role of higher education in teacher training came to include more than just the preparation of teachers. By the turn of the twentieth century, the University of Wisconsin had established a faculty committee that worked with secondary schools in appointing teachers. This committee recommended students to fill vacant teaching positions and further reflected a growing role for higher education in the preparation and placement of teachers in the secondary schools. The committee on the appointment of teachers, which combined with the schools' accreditation committee, sought "to bring forward the best prepared and ablest for positions in the school."[80] Higher education was coming to exert a direct influence over the training and appointment of teachers for secondary classrooms.[81]

# VI

The Committee of Ten represented an attempt to further the role that professional educators and experts played in the secondary schools and in the articulation of these schools with higher education. Eliot's report utilized the nation's educational experts to detail what the secondary schools should teach, and it established a basis for putting the modern subjects on a par with the classics, a step that benefited the secondary schools. New chairs in teaching and schools of education similarly characterized attempts to establish professional control over the content of courses in the high schools and the teaching of those courses. Through schools of education, universities took charge in preparing future teachers and administrators, and they educated new principals and superintendents in their image. The emphasis on highly trained teachers, well-prepared in academic disciplines and in pedagogical skills, reflected the growing stature of education in society and the importance placed on ensuring that students attended classes with competent teachers. University-based training and appointment programs, along with the recommendations embedded in the Committee of Ten, focused prominently on establishing a professional teaching corps and on creating the basis for standardization in curriculum throughout the country.

Eliot hoped that this report would propel the country toward a unified, hierarchical system of education. Although it failed to create such a system on a national level or even to alter many classroom practices, the report underscored a movement toward greater educational uniformity that was occurring in different states and regions. Importantly, it affirmed the vital place of the secondary schools as intermediate institutions that led students from the elementary and grammar grades up into the colleges and universities. One of the report's central purposes had been to stress the importance of secondary education for students going to college and for those ending their education at earlier ages, and Eliot's report clearly sought to encourage more students to remain in secondary school, graduate, and then consider entering college to take up advanced subjects.

The committee and its report emphasized a number of ideas that had dominated and would continue to preoccupy educational debate. Indeed, its lasting influence came by highlighting the ideas that other committees and organizations eventually would use in further articulating American education. What followed in later years was a return to regional initiatives geared toward articulation. As regional efforts intensified in the early 1900s, the relationship between the two educational levels grew even closer. These regional projects—primarily through associations like that established in New England—ultimately led to new attempts to facilitate a national uniformity in America's secondary schools and colleges. The Carnegie Foundation for the Advancement of Teaching was one such notable and successful effort. To move toward national uniformity, this foundation and other education reformers returned to ideas and initiatives embedded in Eliot's report and in the report of the Committee on College Entrance Requirements.

# Chapter 6

# Regional Efforts and a Renewed Focus on National Reform

## I

"What we call the American educational system is composed of a number of separate institutions, each originally built up for some specific purpose and without particular reference to any of the others," Henry S. Pritchett, the former president of the Massachusetts Institute of Technology, announced in 1909. "There must be some way," he continued, "of coupling consecutive stages that will form a vestibuled passage and avoid the confusion and waste of a missed or doubtful connection. In a word, regular temporal succession suggests, in the interest of efficiency and economy, genuine educational continuity."[1] He certainly was not saying anything new. James B. Angell, Charles W. Eliot, and a host of others, in the absence of any governing or central authority, had been trying to move secondary and higher education in this direction since the early 1870s and with notable accomplishments at state and regional levels. Pritchett's aspirations, however, stretched far beyond regional goals. Like Eliot, he wanted national uniformity and a standardized system of education that extended from one coast to the other. What made Pritchett hopeful that success would not be elusive this time, as it had been for Eliot's Committee of Ten, were the vast sums of money he controlled as president of the Carnegie Foundation for the Advancement of Teaching.

From the beginning, Pritchett envisioned using the $10 million gift from Andrew Carnegie, the steel magnate, to do more than disburse pensions to retiring professors. He wanted to reform education. To guarantee that pensions went only to professors of worthy colleges and universities, Pritchett set out to establish a basis for identifying the nation's true institutions for higher education. He understood, after all, that not every school calling itself a college indeed was such an august institution. Since part of identifying a worthy college rested on entrance requirements, he expanded his focus to include secondary education. To assist him in this undertaking, he immediately compiled a board of trustees, with Eliot as chairman and the most prominent university presidents and professors as members. The Carnegie Foundation, therefore, represented another stage in the campaign to place education in the hands of professional educators, and Pritchett used his board and the resources of the Foundation to define both higher and secondary education and, through those definitions, to strengthen the connection between the two educational levels.

To accomplish his goals, Pritchett borrowed heavily from Eliot's Committee of Ten and Augustus F. Nightingale's Committee on College Entrance Requirements. The work of the Carnegie Foundation also benefited from and supported the progress in articulation made after 1900 by the various regional associations of schools and colleges. After the Herculean labors of the Eliot and Nightingale committees in the 1890s, articulation efforts shifted to the regional level, where representatives of colleges and secondary schools banded together in associations to strengthen their relationship. The successful initiatives of these regional associations eventually came together at a national level through the work of the National Association of State Universities and its committee on standards and through the Carnegie Foundation and its driving force in Pritchett. What had been up to the early 1900s mainly state and regional approaches to articulation—with the exception of the Eliot and Nightingale committees—coalesced around these two new organizations and moved toward a renewed national focus. By adopting common definitions for subjects and a standard measurement—the unit—for recognizing the work done in those subjects, the regional associations, with the support and prestige of the two national organizations, began to fulfill the hopes of early education reformers for a strong, efficient system of education. One significant result of their efforts, at a time when more students than ever before were in high school, was to standardize the four-year secondary school, especially in the South, as the predominant, if not quite only, route to college. They focused on ensuring that all students completed a four-year preparatory course of study prior to entering college classrooms. What had been a rare transition from high school to college when Angell became president in Michigan was gradually becoming more common.

Before turning to the National Association of State Universities and to Pritchett's work with the Carnegie Foundation, it is important to understand the strides that the regional associations made in articulating higher and secondary education in the early 1900s. In the Middle States, Nicholas Murray Butler succeeded in launching a central examination board that eventually gained stature beyond New York and helped to develop a uniformity in education across the country. New England's colleges further developed the certificate system, even though the region's prestigious universities embraced Butler's College Entrance Examination Board. The North Central States similarly maintained a focus on accreditation programs and developed a uniform commission that handled school accreditation throughout the region. The South, torn between the Middle States and the Midwest, experimented with a common examination board before abandoning it in favor of a centralized accreditation agency. These regional efforts combined with initiatives on a national scale to further connect the higher and lower branches and to establish a stronger articulated system of education in the United States.

# II

Nicolas Murray Butler played an influential behind-the-scenes role in the Committee of Ten. He hosted the initial meeting of the committee in his apartment and lobbied for Eliot as chair. By the new century, however, he assumed a more direct and influential role in reforming American education. To bring the secondary and higher levels into a closer relationship, he devoted his efforts to the creation of the College Entrance Examination Board. Commonly referred to as the College Board, this organization brought together college and secondary school representatives in the mid-Atlantic states to define and administer common entrance examinations in most of the subjects that comprised the secondary school curriculum. For a number of years at Columbia University, he had tried to persuade his faculty colleagues to support such a common board, but it was not until 1899 that Butler convinced the Association of Colleges and Preparatory Schools of the Middle States and Maryland, a regional association since 1887, to embrace the College Board.[2]

Butler's plan for a central board of examiners was not original. When the New England Association first formed in 1885, secondary school headmasters urged the colleges to consider a joint examining board. Eliot had been at the meeting and, in response to this idea, suggested that a common council made up of colleges and preparatory schools could administer uniform entrance examinations in those subject where agreement existed among colleges.[3] Eliot also made formal proposals for a common examination board in 1894 to the New England Association of Colleges and Preparatory Schools and again in 1896 to a group of secondary school men in New York, although to no avail.[4] He wanted a coalition of colleges to administer uniform tests in all of the subjects contained in the Committee of Ten report, since he understood that such a centralized examination would help to ensure that the secondary schools implemented the committee's recommendations.[5] Had they agreed to these plans, the colleges in essence would have developed uniform entrance standards and examinations, and the secondary schools would have been free of the difficult task of preparing students for different requirements at various colleges.

Butler envisioned a common board of examiners that would fulfill the hopes of these early proposals and eliminate the diversity in college entrance requirements that lingered in American education. Over the last decades of the nineteenth century, education reformers had made progress in easing the transition from secondary schools to colleges and universities, but Butler recognized that improvements still needed to be made. "The time has come, some time since, when public interest requires that the same topics when required for admission by two or more colleges should mean the same thing and be stated in the same way," he said in 1899.[6] Eliot had espoused such goals for the Committee of Ten, and Butler now saw a board of examiners as key to ensuring a widespread uniformity. The College Board's common entrance examinations, he proposed, would encourage colleges to maintain uniform admission standards. Colleges would not have to require the same subjects under Butler's proposal or even concur on the grade or score at which students would "pass" the exam, but they did have to agree on the content for all of the subjects comprising the common exam.[7]

Julius Sachs, a New York principal, applauded Butler's idea for an examination board. He trusted that through such a board the colleges would thoroughly define their admission subjects and the requirements within those subjects. He believed that the College Board would ensure "that half a dozen leading colleges mean, when they record in their catalogues a certain requirement, exactly the same thing, and that the examination papers set will stand as an interpretation—their legitimate interpretation—of that requirement." Such an accomplishment, he claimed in 1899, "will be of invaluable service to us who are obliged to determine in a haphazard fashion in advance what particular interpretation each college may make of its statement."[8] His colleague in Long Branch, New Jersey, however, found little to value in the proposal. "At present," he announced, "I am unable to discover any material advantages that would come to the high school through having a common board of examiners."[9]

Enough college and secondary school representatives agreed with Sachs, and, with Eliot's timely involvement, the Middle States Association unanimously adopted Butler's proposal for a common board of examiners. Eliot's support helped to sway wavering members of the association to vote in favor of the proposal, and Butler credited Eliot with getting the proposal through the association.[10] With this authority to act, Butler moved quickly to organize the College Board. In the first year, twelve colleges joined and accepted the certificates of the board—not to be confused with the certificates of the accreditation program—in lieu of their own entrance examinations. A number of other colleges, although not formally joining the board, agreed to recognize the College Board examinations as optional admission tests. The board held its first set of examinations in June 1901.[11]

In the months between ushering the board into existence and the initial examinations, Butler, as secretary, and the rest of the members of the board had to establish exams in the core academic subjects. To do that, they needed to define the content of those subjects. Rather than establishing their own standards from scratch, they used the standards set by Nightingale's Committee on College Entrance Requirements as the common definitions that would be at the heart of the uniform examinations. Although it tried to retain the same subject standards from year to year, the board periodically revised them. In doing so, the board again followed the lead of earlier committees and turned to professional associations—such as the American Historical Association and the Modern Language Association—as well as to committees representative of secondary schools and colleges.[12] It also adhered to key principles of both Nightingale's committee and the Committee of Ten. As with the previous committees, the College Board felt that the secondary schools should teach all students in the same way, regardless of future destinations. It also believed that students should study a few subjects thoroughly and for a sufficient period of time to get the proper discipline from them.[13]

The College Board additionally resembled these earlier committees by inviting the secondary schools to be actively involved in the Board's work. "The board recognizes that it would be quite as inappropriate for a body composed solely of college professors to decide by a vote questions affecting in an important way the curriculum of the secondary schools as it would be for a body of school-teachers independently to determine questions affecting the college curriculum," argued Thomas Fiske, who became secretary in 1902 when Butler accepted the presidency of Columbia University and, by virtue of that position, chairman of the College Board. "In every important problem that affects the relations between the college and the secondary school," he continued, "the judgment of those who have achieved for themselves eminence in the world of secondary education is at least of equal importance with the judgment of those who have attained similar distinction in the college world."[14]

Accordingly, representatives of the secondary schools, usually principals and headmasters, served on the committees that wrote the examination questions and on those committees that critiqued and revised the exams. These committee assignments were important and facilitated meaningful communication between secondary and higher education around entrance requirements and standards.[15] By most accounts, the colleges and universities listened to the needs and desires of the secondary schools and responded favorably to these concerns. The secondary schools also filled five seats on the overall board, well short of the twenty-two colleges and universities that comprised the board in 1902. They were involved in many aspects of the board's work, even though it was also evident that they were not the chief examiners or the chairman and secretary of the board. The College Board dealt with college admission standards, and, as such, it was run by the colleges, with valuable and essential feedback from the secondary schools.[16]

Although only a few colleges initially joined the Board and only Columbia, Barnard, and New York University agreed to replace their entrance exams with the College Board exams—most simply accepted them as alternatives—the College Board began to solidify its role in developing and administering admission tests throughout the early 1900s. As it did so, it increasingly examined more students and started to spread into New England and even into the Midwest and West. In 1901, however, the Board examined only 973 students in nearly 70 different locations (2 sites were in Europe). Most of the students were from private schools and intended to enter Columbia; few went to Harvard, Yale, or Princeton. The following year, over 1,300 students took the Board's exams at 130 locations.[17] These early years were an inauspicious start for an organization that hoped to encourage uniformity and articulation throughout the country.

Even with Eliot's advocacy and support for the Board, Harvard remained unwilling to accept the results of the College Board examinations. As Harvard's dean said to Butler in 1901, "The experiment you have entered upon is certainly an important one, and its progress will be watched with much interest; but the fact must be recognized that it is as yet only an experiment, and those of us who have taken an active part in efforts for uniformity in admission requirements are fully aware of the difficulties that stand in its way."[18] Unwilling to take a risk on this new venture, Harvard stood aloof. The certificate and accreditation program, moreover, remained popular even as the College Board began its work. Many colleges, especially in the Midwest, often accepted Board examinations from students who could not enter by certificate, but these colleges had little incentive to join the College Board. They focused their resources instead on developing accreditation programs.[19]

Butler could be happy, however, that most of the nation's colleges and universities by 1902 had agreed to accept the Board's tests for admission, although few formally joined the Board or gave up their own entrance examinations. Even Yale relinquished some of its traditional opposition to centralized control and agreed to accept the results of the Board's tests, although only after its faculty had reread and reevaluated the exams. Also in 1902, the Board admitted its first New England colleges—the Massachusetts Institute of Technology, Mount Holyoke College, and Wellesley College—to formal membership.[20] This growing support for the College Board and the spread of a common examination helped to ease the pressures on the secondary schools to prepare students for different colleges. Over the years, as more colleges formally joined the board and the numbers of students taking the exams increased, the pressures on the secondary schools declined even more. By 1910, over 3,700 students took the exams, and that year, more students came from New England than from the Middle States for the first time. Importantly, between the Board's first examinations in 1901 and 1910, some of the nation's most notable colleges joined. Their involvement gave Butler's work added prestige. Eliot eventually succeeded in persuading Harvard to join the College Board, which it did in 1904. Brown followed in 1905, with Yale formally joining in 1909 and Princeton in 1910. However, even by 1910, only twenty-nine colleges had joined the Board.[21]

Although colleges and universities in the South and Midwest accepted the certificates of the Board, they did not join the movement in the early 1900s. The accreditation program in these regions lessened the reliance on entrance exams, and in the South, many colleges, based on their standards and size, were not immediately eligible for membership. The Midwest, however, sent a representative to the Board to speak for the needs of students and schools in the region. Even though only four students taking the Board examinations in 1905 wished to enter the University of Michigan and the University of Chicago, for example, a few hundred students from the Midwest and South hoped to enroll in eastern colleges. Higher and secondary education outside of the eastern regions, therefore, could not ignore the work of the Board, and it was important for the Midwest to participate in the Board's deliberations and serve on the committees that created the entrance tests. "The western secondary schools have to do business with the eastern colleges, and we ought to have members on this Examination Board in order that we may exercise some influence in making the examination questions," the principal of the St. Paul High School argued in 1904. It was a slow process but the College Board was beginning to shift beyond its regional boundaries to become more national in scope and character.[22]

Butler did not overly emphasize this point, but he hoped that the College Board would do more than simply administer uniform entrance examinations; he trusted that it would improve secondary education and college preparatory training. By defining the content of each subject and by giving schools an opportunity to compare the results of their students to the results from other schools, the Board, Butler believed, would encourage schools to improve their teaching and work toward a stronger and common standard for preparation in the secondary schools.[23] By promoting definite standards in each subject, it would provide a basis on which the secondary schools could build and improve their work. The Board never dictated the subjects each college needed to expect for admission, and participation in the College Board did not mean that the colleges had to agree on a set of required subjects. It only provided a mechanism for examining students in those subjects that a college did demand. Pritchett, who often attended meetings of the College Board through his position as president of the Carnegie Foundation, praised this aspect of the Board's work. "In the movement which has gone on in the last decade looking toward the unification of our educational system by the adoption of uniform fair requirements for admission," he stated in 1907, "the College Entrance Examination Board has been the most important factor. The work of this board has become national in its influence."[24]

This uniformity, especially as the Board's exams spread throughout the country, made it easier on the secondary schools to prepare students for college. Those schools that traditionally prepared students for different colleges now hoped that more colleges would accept the exams offered by the Board, thereby relieving the pressure on the schools to prepare students in different Latin texts or English novels so that two or more students could pass the exams of various colleges.[25] Wilson Farrand, the headmaster of New Academy in Newark, New Jersey, and a member of the College Board representing the secondary schools, saw this uniformity in requirements and examinations as "most directly beneficial to secondary schools. For the first time pupils for several different colleges can be prepared to meet the same test," he said.[26]

The Board made other significant contributions to the articulation of higher and secondary education. When it began its work in 1901, it only required that every college desiring to be a member have an entering class of at least fifty members.[27] This standard for admission to the College Board was not high, although it did leave out a number of smaller colleges. In 1907, when it amended its constitution, the College Board significantly redefined the basis on which colleges and universities could join. Butler's board now expected college faculty to have strong academic training "adequate to maintain a high standard of teaching." Moreover, each college needed to maintain a proper proportion between faculty and students, although the Board was vague on what this proper proportion was, other than to say that there needed to be enough professors to allow for specialization. Significantly, it denied membership to any college that established and ran a preparatory department "under the government or instruction of the college faculty." Member colleges further needed to have a sufficient endowment or state support, strong libraries and equipment, and a decent physical plant.[28]

Its most striking change in the requirements for membership, however, dealt with the preparation of entering students. Colleges belonging to the Board had to insist on a high level of preparatory work. "There shall be specifically defined and consistently carried out," the Board declared, "requirements for admission which shall in every case be equivalent to a four-year course in a college-preparatory or high school of good grade, able to prepare its pupils for admission to the colleges already belonging to this Board."[29] By insisting on higher standards for entrance, the College Board provided some backing to the secondary schools. Some colleges, especially those in the South, did not always require their students to graduate from high school before entering college. Indeed, many southern high schools provided only a two- or three-year high school course, and some of the region's colleges even admitted students without this preparation. Competition for students between the secondary schools and higher education could be intense, and the College Board attempted to establish a clear line between the two levels and to support the work of the lower schools. This change in its constitution simultaneously recognized and supported the growing role and stability of the secondary schools in society and in their relation to higher education. The route to college through the high schools had become fairly well established by the early 1900s, which the College Board now recognized, supported, and furthered.[30]

The College Board particularly affected female students and women's colleges. By not making a distinction between female and male students, the College Board exams placed women on the same footing as men and allowed them to compete on the merit of what they knew and could do. The College Board thus was fairly revolutionary at a time when many educators argued that women should be freed from the debilitating effects of studying for exams and admitted only by certificate. The College Board undercut this argument, and women's colleges became prominent members of Butler's organization. Four of the initial twelve members were women's colleges, and M. Carey Thomas, the female president of Bryn

Mawr, served as a vice chair from 1900 to 1902. Measured solely by the number of students taking the exams who hoped to enter all-female colleges and excluding those entering coeducational institutions, women represented a significant number of the Board's examinees. Of the 3,250 students taking the exams in 1908, nearly one-third applied to enter the seven most prestigious female colleges in the country. Smith College alone equaled Harvard in the number of prospective students taking the exam; only Columbia and Cornell counted more students. Opening up the College Board to women's colleges helped these institutions raise their standards as colleges generally were doing and to eliminate their preparatory departments, and it laid the foundation for placing them on an equal footing with male and coeducational institutions. The College Entrance Examination Board, then, not only strengthened education and articulation generally but also notably contributed to female education and women's colleges.[31]

# III

While Yale and Harvard joined the College Board by 1910 and admitted students solely through examinations, most New England colleges remained committed to the certificate program. At the time that the Middle States Association launched the College Board, New England's colleges strengthened the certificate program through the development of the New England College Entrance Certificate Board. Nine colleges banded together in 1902 to establish a uniform organization and process for accrediting secondary schools and admitting students.[32] As with admission examinations, great diversity in the certificate system prevailed. Colleges differed on the information they requested from the secondary schools, the basis for their evaluations of schools, and the conditions under which a school could be dropped from the list.[33] A committee appointed by the Commission of Colleges in New England on Entrance Examinations studied the region's certificate programs and declared that this diversity was inefficient. "Each college has its own method of approving schools, similar to those of others, but without relation to them," the committee reported in 1901. "No college knows except indirectly what schools have been approved by others, or how far the lines of approval are the same." This diversity and lack of coordination weakened "any cumulative effect upon the schools of either the approval or the disapproval of the colleges." The committee further called for "greater care in the examination of schools before approval and united action in dropping from the list schools which send pupils imperfectly prepared."[34]

This new Certificate Board was an attempt to impose some consistency on the certificate system throughout New England and to ensure that the colleges were more careful in their accreditation of secondary schools. This common accreditation board required its member colleges to refuse certificates from any school unless the Certificate Board first passed judgment on and accredited that school. It also established standard accreditation procedures. However, unlike the Midwest, where the universities sent inspectors into the secondary schools to evaluate them, the New England Board maintained the practice of its member colleges and evaluated schools on paper only. Those schools wishing official board recognition submitted detailed information about their schools and courses of study. The board evaluated this information in light of college entrance standards, and it withheld accreditation until it could evaluate the work of students from those schools in the first years of college. Schools had to prove their merit by showing that they could successfully prepare students for higher education. Thus, a student's college record had a direct bearing on whether a school earned accreditation.[35]

Finally, the Certificate Board required that all schools on its accredited list prepare students "for college according to some one of the recognized plans of entering the colleges represented on this board."[36] This stipulation did not mean that the colleges had to embrace a uniform standard in admission requirements. It only meant, as Nathaniel Davis, a professor at Brown University, explained in 1906, that each school had to offer enough subjects "to prepare for some course leading to a degree in some one of the colleges represented on the board, and that, if a subject is to be included among those in which certificates are to be granted, an adequate number of periods shall be assigned to it, and a sufficient amount of apparatus for its proper presentation must be in the possession of the school."[37] This requirement set up some basic and broad standards for the type of preparation expected, but it did not absolutely force the colleges to set uniform admission requirements.

The combination of the Certificate Board with the College Board, however, likely meant that a uniform standard did emerge. The region's colleges, even if they admitted students primarily on certificates, accepted the results of the College Board, especially for those students from nonaccredited schools. Additionally, Harvard and Yale, two of the most popular colleges in the region and country, refused to participate in the Certificate Board or to accept students on certificate, but they did join the College Board and admit students through the Board's examinations. Furthermore, if students in New England wished to enter colleges in other areas of the country, they generally had to take the examinations administered by the College Board. Thus, this region's secondary schools could not ignore the standards of the College Board. Moreover, hundreds of students from other regions took the College Board examinations and sought to enroll in New England's colleges. Although most of them went to Harvard, Yale, and the Massachusetts Institute of Technology or a handful of female institutions, some also went to Brown, Amherst, and other colleges that admitted by certificate.[38]

Consequently, the College Board was a significant factor in New England, even though the Certificate Board remained the dominant method of admitting students. This reality meant that the colleges generally had to develop uniform admission requirements in line with College Board standards, to compete with each other and to draw students from New England's secondary schools and from schools in other regions. This emerging uniformity relieved some of the pressure on the secondary schools to educate students for a number of different colleges and universities. What had been accomplished in the Middle States toward uniformity in admission standards through the College Board also occurred in New England and worked to the advantage of the region's secondary schools, whether they sent students to college through the certificate program or through the examination system.

Just as New England attempted to strengthen its certificate program, colleges and universities in the Midwest took steps to develop a more comprehensive and unified inspection and accreditation program. The guiding force behind this effort was the North Central Association of Colleges and Secondary Schools, a regional association that first brought secondary schools and colleges together in 1895. This organization formed at the impetus of the secondary schools and evolved from a call at the 1894 meeting of the Michigan Schoolmasters' Club. At that meeting, William Butts, principal of the Michigan Military Academy, encouraged his colleagues to form an organization with the colleges to strengthen their relationship with each other and to eliminate the diversity in college entrance requirements. For the first few years of its existence, this organization mainly met socially in annual meetings to listen to papers and debate issues. By 1901, however, it was ready to take a more decisive approach to resolving issues between higher and secondary education, and it embarked on an effort to strengthen and unify the accreditation program that most states had developed independently of each other.[39]

This association, representing schools in the Midwest and some schools in the West, confronted many of the same issues that plagued the certificate program in New England. Not all of the region's colleges and universities agreed on standard entrance requirements or on how best to inspect and accredit schools. This situation became particularly problematic as more and more students crossed state lines to attend college. To meet this need, the University of Michigan, followed by other universities, began to accredit out-of-state schools in 1884. This situation placed the secondary schools in the position of being visited and inspected by a number of different universities. Some universities and colleges had earlier agreed to accept students from schools accredited by other institutions, but not all of the region's colleges supported this approach or universally accepted students accredited by any college or university. The region's colleges, unlike those in New England, consequently spent a great deal of time and money visiting these schools, only to duplicate their efforts with other colleges. As Allen S. Whitney, the principal inspector of schools for the University of Michigan, explained in 1901, "This duplication of inspection is not only a waste of energy and expense; it is a source of annoyance to the schools. The teachers are becoming tired of it. There is danger that the system of inspection will lose prestige in the eyes of the secondary teachers." Hoping to avoid such a problem and to find some relief from the expense and time involved in inspecting schools, the colleges and secondary schools in the association proposed a commission on accreditation to coordinate the region's inspection work.[40]

This commission, approved by the association in 1901, had the task of securing "uniformity in the standards and methods, and economy of labor and expense, in the work of high school inspection." The association further gave the commission the responsibility for annually preparing lists of accredited schools in the North Central states. But the commission did not limit itself merely to issuing lists or coordinating the states' inspection programs; it also focused on clarifying the standards of the secondary schools and sharply dividing the work of secondary education from higher education. The association also asked this commission "to serve as a standing committee on uniformity of admission requirements for the colleges and universities of this Association." With these dramatic steps, the Association pushed forward with an ambitious agenda to coordinate the states' inspection and accreditation programs and to take a more active role in defining the work of the secondary schools and the admission standards of its member colleges.[41]

This commission had the paramount duty "to define and describe unit courses of study in the various subjects of the high school programme, taking for the point of departure the recommendations" of the Committee on College Entrance Requirements. Once again, Nightingale's committee, which rested on the Committee of Ten report, was proving to be influential in ways that neither committee had previously envisioned.[42] Incorporating the work of these committees, the commission on accredited schools in 1902 developed thorough reports and recommendations for each academic subject. These reports detailed what was to be taught in each subject and the number of units assigned to each subject.[43] "A unit course of study," the commission confirmed, "is defined as a course covering a school year of not less than thirty-five weeks, with four or five periods of at least forty-five minutes each per week."[44] This unit system and the development of common requirements for each subject were instrumental in establishing some uniformity in high schools throughout the North Central region.

As reformers often noted, however, this uniformity did not require that all secondary schools offer identical subjects or that colleges demand the same subjects for admission. Indeed, the commission identified only two subjects that it thought all high schools needed to demand of their students and that all colleges should expect for entrance. These absolute requirements were three units in English and two units in mathematics. All other subjects were to be electives that the high schools could offer and the colleges could accept based on their specific needs and traditions. The only other unit expectation was that the high schools would insist on fifteen units for graduation and the colleges would call for the same number of units in their admission standards. This uniformity, then, only meant that two units of American history in Chicago approximated two units in this subject in Denver and not that all schools in the region had to require two units of American history.[45] Eliot had tried to develop this uniformity through the Committee of Ten, but it was left to the North Central Association to further refine and develop the scheme.

The commission also dealt with the region's inspection programs, and its influence on articulation and on the parameters of both higher and secondary education was significant. If schools wished to earn a spot on the association's accredited list, they had to meet a number of criteria that dealt with teacher preparation and responsibilities as well as with the tenor and spirit of the school. The commission trusted that schools wishing to gain accreditation would have adequate libraries, laboratories, and scientific equipment. It also outlined a more ambiguous characteristic of strong, effective schools. "The *esprit de corps,* the efficiency of the instruction, the acquired habits of thought and study, and the general intellectual and ethical tone of the school are of paramount importance," it stressed in 1902.[46]

The commission went beyond this spirit of a school to identify essential teacher qualifications for the secondary schools. Each teacher, the committee detailed, needed to have a "minimum scholastic attainment" that was "the equivalent of graduation from a college belonging to the North Central Association" and "special training in the subjects they teach." It also hoped that new teachers in the secondary schools would have spent time observing skilled teachers and gaining teaching experience as part of their college preparation. Furthermore, no teacher was to teach more than five periods each day. By 1904, it added new requirements and refused to accredit schools that had fewer than five teachers or schools that had more than thirty pupils per teacher, although in recognition of smaller schools, it dropped this minimum number of teachers to four in 1907. Only those schools that ranked well in "these particulars," it concluded, "as evidenced by rigid, thorough-going, sympathetic inspection, should be considered eligible" for inclusion on the accredited list. The Committee of Ten had focused on the need for better-trained teachers, and the North Central Association now outlined specific steps to guarantee that highly qualified teachers filled America's classrooms.[47]

What had started in Michigan as an inspection program that shifted admissions from tests to credentials was becoming a uniform and regional system that rested on clear standards and expectations. The universities gave the secondary schools authority for certifying that students were educated well enough to receive a diploma and continue their education in college. The standards that the North Central Association supported and developed helped to ensure that these students indeed were well qualified.

To identify the schools that met the association's standards, the commission recommended the appointment of a board of five members to evaluate the region's high schools and create a list of schools that deserved accreditation. This board was not to replace the work of state inspectors or impose any standards on universities that might wish to maintain their own inspection programs and accredited lists. Whitney, who presented the report to the Association, explained in 1902 that this board was to create an initial directory of accredited schools that were of "the first rank about whose standing there could not be any doubt in the minds of the authorities of any university represented in the Association." He wanted this record to be "an honor list" prestigious enough to encourage schools to strengthen their courses of study as they worked toward recognition. Such a select list would "assist immeasurably in strengthening secondary education in the Northwest."[48] The association agreed and produced its first accredited list in 1904. There were only 156 secondary schools on it. In comparison, both the University of Michigan and the University of Wisconsin each had over 200 schools on the accredited rosters. The association's standards for accrediting schools were often higher than those of the various state programs.[49]

Since the number of schools accredited by the association was dramatically smaller than the number on many state accreditation lists, the actions of the association were not binding on any institution. A university was not compelled, by virtue of its membership in the association, to reject students from schools that had not earned a place on the association's roster.[50] Even with this qualification, Andrew Draper's support was tepid. Draper, president of the University of Illinois, doubted whether an accreditation program sponsored by the association had any merit. "I suppose there are quite a number here who know that I have been a doubting Thomas about this whole matter," he conceded at the 1903 annual meeting of the association. "What is the real point, what is the educational advantage from this movement in this territory?"[51] Henry Pratt Judson, dean of the University of Chicago, tried to convince Draper to offer heartier support for the accreditation program. "I venture to say that a considerable number [of students] come to Champaign [home to Draper's university] from other states than Illinois." The association's accreditation system, he claimed, would eliminate the time and expense that the University of Illinois had to spend in accrediting schools from out of state, and it would make it easier for students from other states to enter Draper's institution. "If we have a fair degree of unanimity throughout the colleges in the Association, why the colleges in the district know that if a student comes from a certain school they will have no more bother about him," Judson declared. "In other words, it will lessen work." The result would be a much more efficient inspection and accreditation program for all involved.[52]

Judson was such an ardent and enthusiastic supporter that as the association further developed its work in inspecting secondary schools, he saw no reason why it should not also inspect and accredit colleges. "A new line of effort," Judson proposed in 1904, "is suggested by a remark coming from a well-known secondary school man." This eminent educator, according to Judson, claimed that

> the Commission has done a great work in leveling up the secondary schools, in putting a premium on good work and in recognizing the value of inspection. Its attention should now be directed to the colleges. The high schools are being inspected and rated for the benefit of the colleges. Why should not the colleges be inspected and rated for the benefit of the secondary schools and their graduates who are looking for a higher education?

Judson thought that this proposition was "eminently fair," and he could not conceive of any reason why the association should not now focus on the region's colleges.[53] Others agreed with him. "The 193 colleges of our territory are not so superior to the 263 high schools in cities of over 8,000 inhabitants that they need no attention," George Carman, director of the Lewis Institute in Chicago, pointed out in 1906 as the association continued to debate the merits of accrediting colleges.[54] E. L. Coffen, the superintendent in Marshalltown, Iowa, hoped that the accreditation of colleges would eliminate competition with the secondary schools and "either eliminate from the college class certain 'quack' colleges or else cause them to stir themselves to such a degree as to be able to enter the lists of fully recognized institutions." Such inspection and accreditation of colleges, he further argued, would be a great service to parents attempting to choose a college for their children. Since parents often did not know enough to decipher strong colleges from the weaker, "quack" ones, an official list of quality schools was essential, Coffen recommended.[55]

The association moved slowly in developing an accreditation plan for the colleges, but by 1909, it had devised criteria for recognizing colleges and universities. It insisted that faculty have strong academic training and that colleges mandate at least twelve units of collegiate study for a degree. But the association primarily defined higher education by describing it in relation to the secondary schools. "The Standard American College," it outlined in 1909, "is a college with a four years' curriculum with a tendency to differentiate its parts in such a way that the first two years are a continuation of, and a supplement to, the work of secondary instruction as given in the high school." The final two years, it continued, "are shaped more and more distinctly in the direction of special, professional, or university instruction." It further defined a college based on its entrance requirements. "The college shall require for admission not less than fourteen secondary units, as defined by this Association." Fourteen units was an odd standard since the association required high schools to demand fifteen units of study for graduation. A year later it corrected this imbalance and stipulated fifteen units of secondary school preparation as a common admission requirement.[56] As many other organizations were doing throughout the country, this association of schools and colleges embraced the four-year high school as a standard both for a strong secondary school course of study and for admission to college.[57] By 1913, the association had placed seventy-three colleges and universities on its approved list.[58] As Coffen and other educators had hoped, the accreditation of colleges, in addition to that already in place for the secondary schools, was helping to clarify and firmly set the dividing line between higher and secondary education and, thus, effect a strong connection and relationship between them.

In this region, the secondary schools took the lead in building the association of colleges and secondary schools. In the South, a handful of colleges, with Vanderbilt in the forefront, pushed for a regional association, and they hoped to elevate the standards of both higher and secondary education through the association's initiatives. More than in any other region, the South's high schools and colleges—both lacking solid public support—competed with each other for students, and this new association struggled to end this competition, build support for schools, and enunciate a clear division between the two levels. Identifying and enforcing uniform standards for graduation from high school and admission to college were key goals. Therefore, from its beginning in 1895, the Association of Colleges and Preparatory Schools of the Southern States denied membership to any college that had its own preparatory department or admitted students younger than fifteen years of age, and in an important departure from the norm in other associations, it made its actions binding on its members. Consequently, only six colleges were charter members in 1895; ten years later, only eighteen colleges and thirty-five secondary schools had joined. Significantly, most of the secondary schools were private academies and not public high schools, a reflection of the rudimentary state of public education in the South. The association's name and its emphasis on preparatory schools reflected this reality.[59]

The black high schools and colleges in the South's segregated system of education were, of course, left out of this association and its campaign toward articulation. Most white educators in the region likely agreed with Jabez L. M. Curry. An education reformer and member of various northern philanthropic agencies working in the South, Curry argued in 1899 that the focus on white education was appropriate and even essential to the development of African Americans. Although he did not oppose education for African Americans, he believed that there was a "greater need for the education" of white Southerners. "The white people are to be the leaders," he explained, "to take the initiative, to have the directive control in all matters pertaining to civilization and the highest interests of our beloved land." In taking such control, the white man also "will lead out and on other races as far and as fast as their good and their possibilities will justify." This white supremacy, he concluded, was not hostility toward the former slave, "but friendship for him."[60]

With this focus on white education, the Southern Association, led by James Kirkland, the young and energetic president of Vanderbilt University, initially worked to strengthen education by following the College Board's example of a common entrance examination. The association first appointed a committee to study the feasibility of a common system in 1903. As this committee struggled to understand the needs of education in the South and to develop a uniform examination process, it realized that the peculiar circumstances in the region prevented the colleges from adopting the College Entrance Examination Board. The expense of the exams was a hindrance, and the schools in the South were not uniform or consistent enough in their standards and quality of preparation to support Butler's board. More importantly, southern schools were rarely at the level of schools in other regions. The College Board examinations would have been too difficult for most students coming from southern secondary schools.[61]

The committee had considered a certificate board and a uniform inspection program as a way to meet the South's needs, but it maintained that such approaches would be too costly and would consume valuable faculty time. As a result, this committee passed over developments in New England and the North Central states in favor of a regional examination system, in tune with the conditions in the South, as the most efficient way to rectify the situation and improve the state of education. "It would not be possible for this Association to undertake to send a representative to each secondary school," wrote Paul H. Saunders, a University of Mississippi professor who chaired the association's committee on uniform examinations. However, he concluded in 1904, "it is possible to send to each school each year a set of well-selected examination questions." These examination papers, he trusted, would "give to the principal and teachers valuable information as to what results their teaching ought to yield, if it has been successful and proper." Saunders and the rest of the committee members placed great hope in such a system for improving southern education. "Teachers would see wherein their courses were short and defective, and would gradually change their work to correct this," Saunders optimistically predicted.[62]

The first year of this experiment was not a resounding success. The University of North Carolina found the initial tests to be "altogether unsuited to our purposes." One professor at the University of Virginia complained that the examination papers "were not well balanced, if the programmes of our best training schools are a fair test. Some were too easy; others far too long and in some cases too difficult." Saunders sadly realized that "these examinations have not done the work expected of them." Nonetheless, he and his committee were hopeful that the following year would bring more progress, and the association agreed to continue the experiment for another year.[63] Later years failed to show any significant movement in the South toward support for this experiment. Kirkland and Vanderbilt University were the most significant supporters of the program, and Vanderbilt alone accounted for half of all the exams administered. As Saunders pointed out, the examination program was working well for this one institution. The problem was that few other colleges embraced it to the extent that Vanderbilt did.[64]

The examination system lacked widespread support largely because most of the region's colleges admitted their students on certificate. As one college official put it in 1906, "Most of our students are admitted on certificates; we therefore do not find any demand for these examinations."[65] Saunders replied that the exams were crucial to the work of the schools that sent their students to college on certificate. Without examinations, he implied, the accredited schools had no standard on which to judge their work. The "accredited schools above all others need some outside standard of excellence," and the uniform entrance exams, he argued, should be that standard.[66] Nonetheless, the South's colleges and universities never fully embraced a central examination process. In 1912, the committee on uniform examinations formally called on the association to end this short-lived experiment.[67]

As an alternative to the failing examination system, the association created a commission in 1911 to accredit the region's preparatory schools and growing cadre of public high schools. More students than ever before were now enrolling in public schools in the South, and the association dropped the term "preparatory schools" from its title in 1912 and replaced it with "secondary schools" to reflect this new reality. In 1899–1900, the South counted only 1,050 public high schools enrolling just over 61,000 students. A decade later, these figures had changed considerably. Over 2,000 public high schools served 135,000 students in 1911. The number of students from these public schools preparing to enter college also had more than tripled. In the same period, the number of private schools dropped from 755 to 583, even though enrollment in these schools remained stable.[68]

The new commission to accredit these schools established a broad set of parameters in an effort to guarantee that the South's schools improved in quality and not just in quantity. Echoing the standards of other regional organizations, the Southern Association declared that each school had to have at least three teachers with "a college degree from an approved college, or its equivalent," and these teachers had to give all of their time to teaching in the high school. The association wanted to avoid accrediting schools where many of the teachers also taught elementary students part-time. Additionally, each school needed to hold classes for at least thirty-six weeks annually, with a standard forty-minute class period, and the schools needed to provide adequate libraries and laboratories so that students could develop their scientific and research skills. Perhaps most importantly, schools wishing to earn approval by the association had to offer a four-year course of study that included at least fourteen units of study. In 1912, it clarified what this stipulation meant and, in so doing, reflected national trends. "The Commission shall describe and define unit courses of study in the various secondary school programs, based on the recommendation of the Carnegie Foundation and the rules of this Association as herein prescribed." Such a requirement was significant since the two- and three-year high school course was prevalent throughout the South. The South was slowly trying to improve its educational system in line with trends in other regions.[69]

To determine which schools had reached these standards and deserved the recognition of the association, the commission maintained committees in each state that collected and evaluated information on the schools. In gathering this information and determining whether a school met the standards of the association, the state committees relied on personal visits by inspectors, often the ones already connected with universities and doing this work, reports from principals on the preparation of their students, and the records of students in college. Ultimately, the association reviewed these individual lists and compiled its roster of accredited schools from these state reports. By 1915, this roster included 308 schools; five years later it named over 400 schools. It was not until the early 1930s that the association, with the support of the General Education Board, began to evaluate and accredit black high schools and colleges. Before the Association took on the work of inspecting black institutions, black educators banded together to form their own association of secondary schools and colleges and to accredit these institutions.[70]

The Southern Association's accreditation system for white schools, and earlier efforts at implementing a common examination board, helped to align the South with changes occurring in other regions. Its support for a four-year high school based on a standard of fourteen–fifteen units of study matched reforms occurring in other regions—a reform that the Carnegie Foundation attempted to solidify nationally. In the same way that the North Central Association crafted a consistent inspection and accreditation program, the Southern Association devised a set of standards and criteria that helped to unify the region's different accreditation and inspection programs. Although the association grew slowly, it established a higher ideal for education in the southern states, and it encouraged schools throughout the region to strengthen their work. Importantly, it took a stand against preparatory schools associated with colleges. This action ultimately supported the development of public high schools in a region that lacked strong educational opportunities for many students. The South continued to lag behind other areas of the United States in building a strong, unified system of education, but the steps the association took in the first decade and a half of the twentieth century helped to strengthen and standardize education in the South and, gradually, to bring this region into line with schools throughout the country.

# IV

The regional associations in the South and the North Central states, along with the College Board and the New England College Entrance Certificate Board, were all trying to define secondary and higher education and to

strengthen the connection between the two, but they were doing so independently of each other. There were, of course, important exceptions that
created some continuity across regions. These organizations, for example,
often embraced similar standards for the secondary schools. They also
periodically reached out to each other in an attempt to formalize this
movement toward standardization. The Southern Association and the
North Central organization held joint meetings in 1912 to consider stan-
dardizing their entrance requirements and accreditation programs.[71]
Additionally, the College Board expanded beyond New York and the
surrounding states to include New England. As successful as these efforts
were regionally, however, they did not represent a unified, comprehensive
movement toward national uniformity.

Two new organizations in the early 1900s attempted to solidify these
regional initiatives as a national campaign for reform. The Carnegie
Foundation for the Advancement of Teaching and the National Association
of State Universities proved to be more successful in building a system of
education than previous national attempts had been. In part, their success
came from the groundwork established by earlier committees. Both Eliot's
committee and Nightingale's ambitious work had introduced the idea of a
unified system of education and incited prolonged discussion and debate.
The twentieth century also witnessed an even greater explosion in student
population in the secondary schools and in the colleges and universities. In
1911, nearly 1.2 million students attended public and private high schools,
and over 68,000 students graduating that year (out of a graduating class of
136,442) planned to attend college or another form of higher education.
Establishing some sort of uniformity for the increasing number of students
transitioning from one level to the other was even more imperative than it
had been in 1870. Finally, the Carnegie Foundation and the Association of
State Universities benefited from the successful programs that the regional
associations had developed and implemented. Since no federal agency had
the authority to control the development of schools, voluntary and philanthropic organizations filled the void. These two bodies, most notably the
Carnegie Foundation, stepped into this gap and built on earlier efforts
to further the development of a fully articulated system of schools that
transcended state and even regional boundaries to become national in
scope and orientation.[72]

The National Association of State Universities, following the lead of
George MacLean, president of the State University of Iowa, agreed in 1905
to host a national conference with the four regional associations and the
College Entrance Examination Board. The focus of this meeting was to
"plan for inter-relating the work of these respective organizations, and
establishing, preserving and interpreting in common terms standards of

admission to college, whatever be the method or combination of methods or customs."[73] MacLean's main concern was to find a way to unite the different approaches for admitting students and articulating education among the various regions into one coherent and national system. "That there is a tendency and need and a longing for a common, and indeed a national administration, is evident," he explained. A changing society, along with improved transportation and communication networks that facilitated migration across regions, demanded such steps. "The unifying of the republic, the emphasizing of American ideals with a deepening consciousness of our world-wide relations" necessitated the "recognition and development of a national system of education."[74]

The associations, following MacLean's lead, met in August 1906, and promptly passed a series of resolutions aimed at securing the unity MacLean wanted. The first resolution confirmed the work of the New England College Entrance Certificate Board—which had been invited to attend the meeting—and the uniform accreditation efforts of the North Central Association. The conference encouraged all colleges and universities throughout the country to accept students from schools on the lists of both associations. This important step added prestige to both accreditation boards and made it more likely that colleges and universities across the nation, with the exception of those adamantly opposed to admission by certificate, would accept students from schools accredited by these organizations. Access to colleges across state and regional lines, a key element in a national system of education, was enhanced by this action. Moreover, the conference encouraged the two regional associations that did not have certificate boards or accreditation programs—the Southern Association and the Middle States Association—to develop such unified efforts. The Middle States Association formed a committee to consider whether it should implement a common certificate and accreditation board, but the South eventually created its own commission in 1911 to accredit secondary schools. Beyond ensuring the expansion of the certificate system, the conference recognized and approved of the subject definitions and standards that the College Board had utilized in crafting its entrance examinations. The conference proposed that these definitions, in turn based on earlier national committees, serve as the basis for education across the country. It encouraged the regional associations to join with the College Board "in formulating and revising, when desirable, these definitions." To build on these steps, the representatives in attendance agreed to form a permanent national conference on standards for colleges and universities, and to meet regularly.[75]

The conference in the next few years built on these early efforts toward uniformity by further defining the academic subjects. In 1909, the members embraced the "unit" definition that the North Central states had first adopted in 1902 and that the Carnegie Foundation promoted in the early 1900s. "A unit represents a year's study in any subject in a secondary school, constituting approximately a quarter of a full year's work," it stated. This definition of a unit rested on a number of crucial assumptions that the conference spelled out in detail. As with the other associations, the conference assumed that the four-year high school course was standard and that each student would pursue a course for four or five periods each week for thirty-six–forty weeks. While the North Central Association settled on at least forty-five minutes as an appropriate length for each period, the national conference provided a variable range from forty to sixty minutes. Following this action in the national conference, the College Board and the Middle States Association accepted the standard unit definition by 1910, and the Southern Association did so by 1912.[76] The national conference, working with the regional associations, was providing a way to achieve national consistency in standards and academic preparation both for graduation from secondary schools and for admission to colleges and universities.

By the early 1900s, it was evident that various regional organizations and national associations had agreed on the broad parameters of higher and secondary education. The Carnegie Foundation for the Advancement of Teaching, with Henry Pritchett at the helm from the beginning in 1905, provided crucial support in ensuring that the colleges embraced these standards and encouraged the lower schools to strengthen secondary preparation. Although the ostensible purpose of the Foundation was to disburse pensions to retired college professors, it had to establish criteria for identifying valid colleges. Accordingly, it directly addressed one of the major issues of the day: what differentiated a college from a secondary school, and where did "colleges" that offered primarily preparatory training and elementary subjects fit in an educational system? Initially, it excluded professors from state universities and from overtly sectarian colleges from receiving pensions, but its contribution to setting standard criteria for higher education affected these institutions and influenced the development of secondary education. By defining colleges and universities, in part, based on their relationship to the secondary schools, it also had to provide recognizable characteristics for these lower schools. As Pritchett outlined in 1908, "The first step was to adopt an arbitrary definition of a college, based on qualifications of its teachers and its curriculum; but in order to complete the statement that college should rest upon the high school it was necessary, also, to define the high school."[77]

A core requirement for recognition by the Foundation was that colleges had to require four years of secondary school as standard preparation, among other criteria. "An institution to be ranked as a college," Pritchett announced in 1906, "must have at least six (6) professors giving their entire time to college and university work, a course of four full years in liberal arts and sciences, and should require for admission, not less than the usual four years of academic or high school preparation, or its equivalent, in addition to the pre-academic or grammar school studies."[78] To articulate clearly what a four-year high school should be, the Foundation borrowed from the North Central Association's definition of secondary school units and the subject definitions of the Committee on College Entrance Requirements—which the College Board had adopted and which, in turn, the National Association of State Universities was promoting. The Foundation concluded that the four-year high school was equivalent to fourteen units of study, as long as those units were based on the subject definitions established by these earlier associations. "The better high schools," Pritchett explained, "require pupils to recite on the average four studies daily five times a week. Assuming a study pursued for one year with recitations five times weekly as a unit, the ordinary high school course would therefore furnish in four years sixteen such units." After allowing time for review and for students who changed their minds about the courses of study they wished to pursue, Pritchett and the Foundation concluded that "fourteen such units seem a fair measure of the work of the high school." A college accepting fewer than fourteen units, he declared, likely was "devoting part of the time of its collegiate classes to instruction in subjects which, in a well organized educational system, are now left to the high schools."[79]

Pritchett thought that these requirements were absolutely essential to the future of education. By defining higher education in relation to the number of years of preparatory work, Pritchett hoped to eliminate college competition and give the secondary schools space to grow and thrive. It was important for the state of education, he argued, that the high schools have a chance to improve their work. After all, the colleges and universities would only do better work if they rested on a foundation of solid preparatory schools. Pritchett knew that if colleges had lower standards than graduation from a four-year high school, students would be compelled to leave school early and enter college. He understood that "it is not easy to keep in the upper classes of the high schools boys who are free to try their luck at the university." The colleges then would have to devote most of their resources to preparatory training. Consequently, Pritchett warned that any college offering secondary school work as part of its collegiate courses would not be eligible for the Foundation's pensions. Pritchett recognized that the high schools would never succeed and flourish as long as they had to compete with the colleges and lost students to them. "Unless the college is to articulate with the high school, the system of education in any community cannot be a consistent one" or a strong one able to meet the needs of society, he implied. Pritchett used the Foundation to distinguish the work of each educational level and to build a harmonious relationship between the two.[80]

This need to clearly define the responsibilities of the two levels was particularly crucial in the South. Although the Foundation identified fifty-two institutions in 1906 that deserved recognition, few were in the southern region. Not surprisingly, over half of the approved colleges and universities were in New England, New York, and Pennsylvania. Almost none were from the South and only thirteen came from western states. "It was inevitable that any choice of institutions which took account of educational standards or denominational limitations, and which excluded state institutions, should have some such result upon its first application," Pritchett argued in 1906. Many of the schools in the southern and western regions were either denominational or state-sponsored. As such, they were not eligible for Carnegie support, until the Foundation accepted state universities in 1908. As Pritchett made clear, however, many of the southern colleges did not meet the criteria that the Foundation had specified. "In the south very few institutions require of their students conditions of admission such as are enforced in all colleges upon the 'accepted list,'" he said. Vanderbilt had high enough standards, and Tulane University was moving in that direction; most colleges, however, "even those of age and high standing," did not enforce "entrance requirements which made any sharp distinction between the high school and the college."[81]

Pritchett hoped that the southern universities would embrace the standards of the Carnegie Foundation and pointedly distinguish between secondary and higher education. President Kirkland of Vanderbilt thought that the Foundation was having an effect. "When we remember how this Association [of southern schools] has struggled for fifteen years for the promotion of better standards, and how difficult, if not impossible, it has been to secure advancement," he said, "we cannot withhold an expression of surp[r]ise at the readiness with which old standards have been abandoned and new ones adopted in the presence of the simulating influence of the Carnegie report."[82] Nonetheless, in 1910 when the Foundation listed seventy-one approved schools, only five were in the South, including three in Missouri; nearly twenty-four were in western states.[83]

Pritchett's great hope was that the Foundation would be national in scope and articulate education in the South and across the country. This ambition frightened a number of critics, who feared that a centralized agency would direct the schools and eliminate local control. Ohioans were particularly angry, since Pritchett denied pensions to Ohio's state universities.

Carnegie had expanded his gift in 1908 to include state schools, but Pritchett argued in 1910 that Ohio's system of colleges and universities was inefficient and unworthy of recognition. An Ohio newspaper reacted angrily to this news and criticized the Foundation for trying to "Pritchettize" education or create schools and institutions that looked the same in each state. "To what extent," the newspaper asked, "are the states of this Union to submit to a private overlordship of their public educational systems, in order to obtain a few paltry pensions for superannuated college professors?" The newspaper continued to argue that "the pretensions of Dr. Pritchett at present suggest that he would soon develop an educational primacy and authority in effect scarcely inferior to that of the minister of education in the kingdom of Prussia." A high school teacher similarly complained that the Carnegie Foundation basically took control of the nation's schools and that college presidents had to have Pritchett's approval before they could admit students.[84]

As these criticisms highlight, Pritchett's efforts engendered controversy, but so had many of the earlier reforms and ideas promoted by Angell, Eliot, and Harper. Pritchett's sin, at least in the eyes of his critics, was to lead a foundation that had the resources and reach to standardize an educational system throughout the nation. Carnegie had amassed a fortune by building efficient, complex organizations that had then shaped the nation in profound ways. It was perhaps inevitable that his foundation would encourage such development in education. Still, the Carnegie Foundation was only doing on a national scale what many other organizations had been doing at state and regional levels. Since no centralized agency existed to direct educational matters, Pritchett offered the Carnegie Foundation to fill this need. It provided leadership in shaping a finely articulated system of education for the nation, but it was an organization that rested on the efforts of other reformers.

In response to criticisms, Pritchett maintained, as Eliot had before him, that local needs and traditions could still influence the curriculum of the schools and universities. "To bring the requirements for admission to college approximately to the same level throughout the country does not involve," he asserted in 1907 and continued to proclaim thereafter, "any procrustean system of college education. One college may insist upon a definite group of subjects a prerequisite to its courses and another college may insist upon an entirely different group, but the two colleges by their dissimilar requirements may still be demanding of their candidates for admission an equal amount of mental development." It was this equivalence of development that Pritchett wanted from education and not some curriculum that looked the identical from one state to another. "Carnegie unit, as it has been called, does not undertake in any way to fix the program of secondary education," he declared. "It has absolutely nothing to do with what a secondary school may teach or what a college may require." It only provided a way to measure the work of one school and compare it to that of another. Variation among institutions was important and a hallmark of American education, Pritchett believed. However, "this variation is no excuse for the difference in the amount of academic training required before the college work is taken up," he concluded.[85]

The national scope of the Carnegie Foundation's efforts, he told his critics in 1907, was what made the organization so important and necessary. "It is, in my judgment," he proclaimed, "a wholesome influence in education to have a few such centralizing influences" that would unify education throughout the country. "Our tendencies in the past in the founding and maintenance of colleges have been almost wholly along competitive lines." No agency has existed to encourage these institutions to think broadly and nationally about education. "Here for the first time," he boasted, "is created an agency which is conscientiously seeking to consider the problems of institutions from the larger view of the welfare of the teachers in all colleges and universities, and to take into account the interests not alone of a community or of a section, but of a continent."[86]

In the early decades of the twentieth century, the Carnegie Foundation built on state and regional initiatives—some in existence since the 1870s—to become a national agency that furthered the establishment of an efficient system of articulated schools and colleges across the continent. Had he been alive in the first decades of the new century, James McCosh would have witnessed an important stage in the development of a uniform stairway connecting schools and colleges throughout the country, and he likely would have applauded the influence of Andrew Carnegie, a fellow Scot, and his foundation.

# Epilogue

## Looking Ahead by Looking to the Past

### I

Toward the end of his address to the National Education Association (NEA), James Angell told a humorous story that aptly illustrated what was then a pressing issue in education. The theme of his story was "the open season for abusing the colleges," and it started with a familiar refrain: "Somebody was abusing the colleges" and claiming that they were not doing a good job educating students, "and it occurred to a college professor, being of an inquiring turn of mind, to follow this up." This professor "found, of course, that all his college colleagues said that the trouble was with the high schools, that they were certainly rotten institutions, and they sent a very poor type of young person up to the college." In turn, the professor "went to the high schools and inquired of the teachers and principals what about it. They said there was some, though not much, truth in it," But if he would go to the lower grades "and see the young savages who came out of the grades into the high school and then saw what the high school was asked to do in four years, he would see how impossible it was to send on much better stuff to the colleges."

The story continued in this way, until the trail led the professor to a kindergarten and eventually to a home, where he spoke with a young pupil's mother. And what did the mother say, he asked? She knew that her child was bad, although not as terrible as most of the neighborhood children. Anyway, she concluded, it was all his father's fault.

To make the point explicitly, Angell told his audience that the universities and high schools needed to stop playing "the great American game of passing the buck" and start working out a better relationship with each other.[1]

The speaker who provided a light moment in an otherwise sober annual meeting, however, was not James B. Angell, the president of the University of the Michigan, but James R. Angell, the president of Yale University and the son of the former Michigan president. In 1928, when he stood before his NEA audience, he argued, much as his father had half a century before, that the nation needed a stronger, more efficient system of articulated schools.

The younger Angell's story was as familiar in 1928 as it was when his father first took steps in the 1870s to establish a stronger, harmonious relationship between higher and secondary education. And, it is a common refrain in yet another century. In the twenty-first century, would-be education reformers have continued to highlight the lack of a strong connection between the higher and secondary branches of American education. Bill Gates, chairman of Microsoft and cofounder of the Bill and Melinda Gates Foundation, argued in early 2005 that "America's high schools are obsolete" and were "limiting—even ruining—the lives of millions of Americans every year." Not only that, they were threatening the health of the nation by failing to prepare students for college, work, and citizenship. These schools needed to be drastically reconsidered and altered, Gates argued, so that "every kid can graduate ready for college."[2] As Gates pointed out, the high schools continue to play crucial roles in America's educational system.

Even before Gates took an interest in education, politicians, scholars, and policymakers had searched for ways to forge stronger links between higher and secondary education in the latter decades of the twentieth century and the early years of the twenty-first. Today, throughout the country, advocates of K-16 and PK-20 initiatives seek to ease the transition from grade to grade and from one school to another at all levels of education. They envision a seamless system that stretches from the kindergarten through the elementary and high school grades and into college, or even from the preschool years into graduate and professional school. Some of these initiatives claim university backing and others are housed in education policy centers, but all hope to bring more students into higher education so that they and the nation can benefit from advanced education.

Education reformers have discussed the crucial need to link higher and secondary education in three different centuries. Does this fact mean that Michigan's president—or Charles W. Eliot, Augustus F. Nightingale, William T. Harris, Cecil Bancroft, and a host of other reformers—had no noticeable effect on the relationship between the nation's high schools and colleges? The historical record suggests that they had a pronounced influence and had altered the shape of education significantly between the time when James McCosh first regaled his NEA audience with tales of woe and Henry Pritchett directed the Carnegie Foundation for the Advancement of Teaching.

# II

Throughout the last decades of the nineteenth century and the early years of the twentieth century, reform-minded educators made tremendous strides in articulating higher and secondary education. From the accreditation programs prevalent in the Midwest and West to the College Board in New England and the Middle States and to the hesitant steps in the South, reformers succeeded in bringing the two educational levels closer together. In the process, they began to clearly define the work of secondary education and to distinguish it from the higher work offered in the colleges and universities. They solidified the four-year high school as the standard route to college, strengthened the place of secondary schools and higher education in society, and underscored the role of expertise and experts in the life of the nation. As a result, more students were leaving high school and entering college, which was a boon for higher education and for those who believed that a college education would funnel students into the middle-class positions desperately needed by the nation. What had been a loose grouping of schools coalesced into a tightly organized system of education with a curriculum that looked remarkably similar in Michigan, Massachusetts, and even Louisiana.

Reformers, nonetheless, had not succeeded completely in building a fully articulated system, and problems persisted. Complaints by high school teachers about the unreasonable demands of higher education remained a common anthem in regional and national meetings, and many "colleges" still did not deserve that name, based on a common definition of higher education evolving through the work of foundations and regional associations. These institutions enticed students into their classrooms and hindered attempts to create a sharp division between colleges and secondary schools. However, whether they were building a stairway or grading a broad avenue that led from the lower to the higher branches, reformers representing both higher and secondary education clearly had made progress at the turn of the twentieth century in creating an efficient system of education.

Getting to this point, where Henry Pritchett could extol the virtues of articulation on a national level, had not been easy, and further strengthening the connection between the two educational levels today remains challenging. Creating a seamless educational system was difficult in part because the two educational levels evolved throughout the nineteenth century with different traditions, missions, and cultures. This variation in missions was something Michigan's president wrestled with when he launched one of the nation's first attempts to create a system that linked his state's high schools with the university. James Kirkland, Oscar Robinson, John Tetlow, Henry Pritchett, and many other reformers similarly struggled with this challenge in the last decades of the nineteenth century and the first years of a new century.

The high schools—or the "people's colleges," as they often were called—provided an emerging middle class throughout much of the nineteenth century with a way to maintain and enjoy that social class status. By the turn of the twentieth century, however, the colleges and universities began to argue that their role was to ensure that the burgeoning middle class retained its professional standing in a nation in the throes of industrialization and enjoying striking scientific advances. Higher education and not the secondary schools, college presidents and faculty proclaimed, had the expert professors capable of navigating the transition to an urban, industrial nation and training a new generation of skilled professionals to tackle the problems facing the country. To fulfill this mission, colleges and universities needed to offer more advanced and rigorous courses than they had been doing for much of their history, and this meant ensuring that a strong college preparatory focus existed in the secondary schools.

Educators in the nation's colleges and universities thus argued that the secondary schools needed to do a better job of preparing students for the demands of college, which led the high schools to complain that the universities were undermining their traditional mission. For teachers and administrators in the secondary schools, the high school course was complete in itself and prepared students in the modern subjects of English literature, history, mathematics, sciences, modern languages, and geography for the demands of life. The secondary course did not need to be capped with a college education, they claimed. As the people's colleges, the high schools long had provided the skills, cultural polish, and refinement that led to positions in the middle class.[3] The colleges, they complained, wanted the secondary schools to deemphasize a broad focus on the modern subjects and the needs of life to prepare students more narrowly for college where they would then finish their education. The transformation of society to a more urban and industrial nation, however, led the universities to argue that they could do a better job preparing America's young for the future.

At the heart of the tension between the high schools and the universities, then, was a debate over which institution would do a better job of educating students for middle-class positions that would meet the demands of society. The universities wanted this responsibility. The secondary schools understandably were loath to give it up.

In taking on this responsibility, the colleges and universities moved in the direction of the lower schools by embracing degree programs and admission requirements that paralleled many of the modern courses offered in the high schools—although not always in ways that directly aligned with the work of the lower schools—but the secondary schools remained leery of increasingly bold attempts by the nation's colleges and universities to encroach on their mission. They were losing the battle. The high schools could go only so far in building on the elementary course, and the universities were there to pick up where the high schools stopped. By aligning more closely with the lower schools and building on the work of the high schools, the universities were opening their doors to more students and solidifying their place at the top of the educational pyramid. These actions represented a declining place in the educational hierarchy for the high schools, even as the secondary schools, by awarding the high school diploma, became the gatekeepers to higher education. The frustrations of the high schools and confusion over their role extended into the following decades, and the high schools continued to wrestle with their role as secondary schools in society and in relation to the nation's institutions of higher education. Great strides had been made in creating an educational ladder, but differences in purpose and confusion over roles remained.

Articulating higher and secondary education has been challenging, and the historic difference in missions between the two institutions is part of the reason. Yet, the story is not just one of challenge but one of success in bringing the two levels into greater harmony. Leaders and policymakers today continue to bemoan a lack of alignment between higher and secondary education, but this history of articulation underscores the initiatives that succeeded in strengthening the connection between the two educational levels and that set the foundation for efforts that continue today. The most successful of these early reform efforts shared characteristics that current policymakers might wish to consider.

The reforms that had the most effect were those backed up with incentives and that insisted on some sort of accountability, in the language of today's education policymakers. The College Board in the early 1900s promoted specific content for each subject and tested students based on that knowledge. Initially, at least, the content of the subjects that the College Board tested rested on the work of the Committee of Ten and the Committee on College Entrance Requirements. These committees, when they had completed their work in the 1890s, did not lead to a change in high school courses of study. Schools had little incentive to embrace the reforms these committees recommended. But, in combination with the College Board and the access such examinations provided to college, the secondary schools had a reason to change their requirements. Moreover, if students from one school repeatedly failed to pass the College Board exams, those students and their schools gained an unfavorable reputation. The Carnegie Foundation backed up its calls for reform with pensions for professors, which increased the pressure on the universities and in turn on the high schools to establish verifiable standards and strengthen their relationship with each other. Earlier in the 1870s and 1880s, the University of Michigan and other colleges throughout the country developed programs that similarly exerted pressure on the lower schools but provided a crucial incentive. They eased the way for students to enter the university and made it easier for schools to prepare students to meet college admission standards. However, if students from a particular school failed, the colleges threatened to remove that school from the accredited list. This action reflected directly on the principals. Higher education meanwhile had an incentive to embrace the modern subjects of the high schools. Many of the colleges needed students and they did not want to shut themselves off from the steady supply that the high schools could send them.

Further, important reforms came out of professional organizations where educators from both levels came together to discuss crucial issues and reflect on their work. Some of these were mainly social organizations that brought people together to share ideas. For a number of years, the annual meetings of the National Education Association were geared more toward discussion and debate than policy actions. But many meetings and organizations that included representatives from higher and secondary education led to specific and direct actions that altered the shape of education. The North Central Association and its focus on standard accreditation procedures owed its existence to William Butts, the principal of the Michigan Military Academy, and the Michigan Schoolmasters' Club, where Butts sponsored a resolution calling for the creation of a regional association of schools and colleges. The College Board combined secondary educators with college professors and presidents, and, as with the North Central Association, these two groups of educators shared and discussed crucial issues. They formed a dynamic, cooperative team that then worked together to implement their ideas in America's schools and colleges. These professional organizations provided their participants with opportunities to reflect on their work, share ideas, and consider how to build a strong educational system, and they then provided the support and resources to implement those reforms. In the absence of a central authority to control America's schools, these professional associations provided crucial support in organizing the nation's schools into a system of education, and they did so by bringing representatives from the two levels together.

# III

That Yale's president emphasized the need for articulation in 1928 and that a powerful entrepreneur and philanthropist made similar arguments nearly eighty years later does not mean that the first efforts at articulation in the 1870s failed. Contexts shift and the demands on education today reflect new social and economic realities, just as they did at the turn of the twentieth century and in the years immediately before the stock market crash in 1929. For example, schools and colleges today enroll more students than most reformers could have dreamed of or hoped for in the nineteenth century, and today's schools are mass institutions with different challenges and expectations. Additionally, access and retention of minority students is a much more pressing concern today than it was in the late nineteenth century, although even then some educators, especially black educators in the South, wrestled with improving high schools for black students and coordinating their connection to colleges.

Nonetheless, as they struggle to adapt education to the changing demands of society, reformers today have much they can learn from these earlier educators who sought to make education a living, dynamic force in the country. Many of the questions that educators faced in the late nineteenth and early twentieth centuries are questions that remain vital today: what are the goals of education, who has access to college, how well are students making the transition from high school to college, and how well prepared are they to succeed in college? Revisiting and analyzing the relationship that emerged between higher and secondary education at the turn of the twentieth century and how educators answered these questions then can suggest possible ways of thinking about these issues and help frame alternative solutions today. It can illuminate the boundaries or barriers that hamper institutional cooperation and clarify the dilemmas that confront today's researchers, administrators, teachers, and policymakers.

Historical research can provide us with a better understanding of what earlier educators and reformers did well, where they stumbled, how they overcame opposition, what compromises they made, and how their decisions continue to affect schools and education today. Certainly, there are differences between the articulation campaign at the turn of the twentieth century and the policies aimed at bridging the gap between high schools and universities today. But, by understanding how higher and secondary education shaped each other and established a system of education in an era that radically changed American society, researchers and policymakers gain a unique perspective on how to ensure that today's system of education is as strong as it can be.

James McCosh, Charles W. Eliot, Augustus. F. Nightingale, and other educators from previous centuries have a lot to offer today's education reformers and policymakers.

# Notes

## Introduction

1. University of Michigan, *The President's Report to the Board of Regents for the Year Ending June 30, 1872* (Ann Arbor: Published by the University, 1872), 9.
2. Charles K. Backus, "President's Report, Detroit Schools," in *Thirty-Seventh Annual Report of the Superintendent of Public Instruction of the State of Michigan* (Lansing: W.S. George & Co., 1874), 313.
3. See, e.g., Richard Hofstadter and C. DeWitt Hardy, *The Development and Scope of Higher Education in the United States*, 3rd ed. (New York: Columbia University Press, 1958); Lawrence Cremin, *The Transformation of the School: Progressivism in American Education, 1876–1957* (New York: Vintage Books, 1961); Laurence Veysey, *The Emergence of the American University* (Chicago: The University of Chicago Press, 1965); David B. Tyack, *The One Best System: A History of American Urban Education* (Cambridge: Harvard University Press, 1974); Robert Church and Michael Sedlak, *Education in the United States: An Interpretive History* (New York: Free Press, 1976); Larry Cuban, *How Teachers Taught: Constancy and Change in American Classrooms, 1890–1980* (New York: Longman, 1984); Roger Geiger, *To Advance Knowledge: The Growth of American Research Universities, 1900–1940* (New York: Oxford University Press, 1986); Daniel Tanner and Laurel Tanner, *History of the School Curriculum* (New York: Macmillan Publishing Company, 1990); Christopher J. Lucas, *American Higher Education: A History* (New York: St. Martin's Griffin, 1994); Herbert Kliebard, *The Struggle for the American Curriculum, 1893–1958* (New York: Routledge, 1995); Julie A. Reuben, *The Making of the Modern University: Intellectual Transformation and the Marginalization of Morality* (Chicago: The University of Chicago Press, 1996); Diane Ravitch, *Left Back: A Century of Battles Over School Reform* (New York: Touchstone, 2000); Herbert Kliebard, "A Century of Growing Antagonism in High School–College Relations," chapter in *Changing Course: American Curriculum Reform in the 20th Century* (New York: Teachers College Press, 2002), 50–60; John R. Thelin, *A History of American Higher Education* (Baltimore: The Johns Hopkins University Press, 2004).

4. Edward A. Krug, *The Shaping of the American High School, 1880–1920* (Madison: The University of Wisconsin Press, 1969).

5. Harold Wechsler, *The Qualified Student: A History of Selective College Admission in America* (New York: John Wiley & Sons, 1977). See, e.g., John S. Brubacher and Willis Rudy, *Higher Education in Transition: A History of American Colleges and Universities, 1636–1968* (New York: Harper & Row, 1968); Frederick Rudolph, *The American College and University: A History* (New York: Alfred A. Knopf, 1968); Geraldine Jonçich Clifford, *"Equally in View": The University of California, Its Women, and the Schools* (Berkeley: Center for Studies in Higher Education and Institute for Governmental Affairs, University of California, Berkeley, 1995); Paul Westmeyer, *An Analytical History of American Higher Education* (Springfield, Ill: C. C. Thomas, 1997); Arthur Cohen, *The Shaping of American Higher Education: Emergence and Growth of the Contemporary System* (San Francisco: Jossey-Bass Publishers, 1998).

6. Nicholas Murray Butler, "Discussion," in *The Addresses and Journal of Proceedings of the National Educational Association, Session of the Year 1891* (Published by the NEA, 1891), 502.

# 1  Changing Expectations for American Education

1. James McCosh, "Upper Schools," in *The Addresses and Journal of Proceedings of the National Educational Association, Session of the Year 1873* (The National Educational Association, 1873), 23.

2. Ibid., 35. For his description of McCosh, see D. C. Gilman, "The Future of American Colleges and Universities," *The Atlantic Monthly* 78 (August 1896), 175–176.

3. S. H. Carpenter, "The Relation of the Different Educational Institutions of the State," *Wisconsin Journal of Education* 4 (March 1874), 86.

4. United States Bureau of Education, *Report of the Commissioner of Education for the Year 1876* (Washington, D.C.: Government Printing Office, 1878), lxxi; *Report of the Commissioner of Education for the Year 1878* (Washington, D.C.: Government Printing Office, 1880), lxxx–lxxxi.

5. United States Bureau of Education, *Report of the Commissioner of Education for the Year 1873* (Washington, D.C.: Government Printing Office, 1874), lii.

6. Ibid., xlvii.

7. United States Bureau of Education *Report of the Commissioner of Education for the Year 1881* (Washington, D.C.: Government Printing Office, 1883), cxl; *Report of the Commissioner of Education for the Year 1885–86* (Washington, D.C.: Government Printing Office, 1887), 362; *Report of the Commissioner of Education for the Year 1887–88* (Washington, D.C.: Government Printing Office, 1889), 494; *Report of the Commissioner of Education for the Year 1890–91* (Washington, D.C.: Government Printing Office, 1894), 789–796, 817–819,

832–833. The historian Edward Krug has argued that scholars should be cautious when using Bureau of Education statistics. As he warned, the statistics for secondary schools before 1889–90 and for higher education before 1900 are unreliable, confusing, and contradictory. They are based on voluntary reports from schools. They are compiled from a number of confusing tables spread throughout the commissioner's reports and likely fail to count precisely the number of students enrolled in secondary and higher education. See Edward A. Krug, *The Shaping of the American High School, 1880–1920* (Madison: The University of Wisconsin Press, 1969), 451.

8. United States Bureau of Education, *Report of the Commissioner of Education, 1873*, lix, 663–682; *Report of the Commissioner of Education, 1878*, xc, 515–546; *Report of the Commissioner of Education, 1890–91*, 817–819 832–833; *Report of the Commissioner of Education, 1885–86*, 490–493. See also Geraldine Jonçich Clifford, *"Equally in View": The University of California, Its Women, and the Schools* (Berkeley: Center for Studies in Higher Education and Institute for Governmental Affairs, University of California, Berkeley, 1995), 3–4. As historians have warned, statistics for higher education in this period are not entirely reliable. See Krug, *Shaping of the American High School*, 451.

9. Robert Wiebe, *The Search for Order, 1877–1920* (New York: Hill and Wang, 1967), xiii–xiv, 1–4, 10–16, 22, 30–39, 44–47, 111; Kenneth T. Jackson, *Crabgrass Frontier: The Suburbanization of the United States* (New York: Oxford University Press, 1985), 20, 272–275.

10. Wiebe, *Search for Order*, 10–16, 22, 30–39, 111–129; Samuel Hays, *The Response to Industrialism: 1885–1914* (Chicago: The University of Chicago Press, 1957), 2–3, 48, 71–75, 188–190; Jackson, *Crabgrass Frontier*, 272–275.

11. Alba M. Edwards, *Population: Comparative Occupation Statistics for the United States, 1870 to 1940; Sixteenth Census of the United States, 1940* (Washington, D. C., 1943), 11. See also Maury Klein and Harvey Kantor, *Prisoners of Progress: American Industrial Cities, 1850–1920* (New York: Macmillan Publishing Co., Inc., 1976), 73–77.

12. William J. Reese, *The Origins of the American High School* (New Haven: Yale University Press, 1995), xvi–xvii, 167–169, 238–239; David F. Labaree, *The Making of an American High School: The Credentials Market and the Central High School of Philadelphia, 1838–1939* (New Haven: Yale University Press, 1988), 4–5, 9–10, 33–34, 46–47, 114.

13. Reese, *Origins of the American High School*, 107–121; Labaree, *Making of an American High School*, 4–5, 9–10, 19–23, 29. See also John L. Rury, *Education and Social Change: Themes in the History of American Schooling* (Mahwah, New Jersey: Lawrence Erlbaum Associates, 2002), 82–90.

14. Laurence Veysey, *The Emergence of the American University* (Chicago: The University of Chicago Press, 1965), 264–66; Labaree, *Making of an American High School*, 6, 65–66, 128–130, 134–135, 165–170; Burton J. Bledstein, *The Culture of Professionalism: The Middle Class and the Development of Higher Education in America* (New York: W. W. Norton & Company, Inc., 1976), 33–39; John R. Thelin, *A History of American Higher Education* (Baltimore: The Johns Hopkins University Press, 2004), 155.

15. C. W. Parmenter, "Discussion," in *Sixty-Second Annual Meeting of the American Institute of Instruction, Lectures, Discussions, and Proceedings, Bethlehem, N.H., July 6–9, 1891* (Boston: American Institute of Instruction, 1891), 76–77.

16. Olivier Zunz, *Why the American Century?* (Chicago: The University of Chicago Press, 1998), 9.

17. There is a significant literature on universities, expertise, and science. Edward Shils, "The Order of Learning in the United States: The Ascendancy of the University," in *The Organization of Knowledge in Modern America, 1860–1920,* edited by Alexandra Oleson and John Voss (Baltimore: The Johns Hopkins University Press, 1979), 19–29; Richard Hofstadter, "The Revolution in Higher Education," in *Paths of American Thought,* edited by Arthur M. Schlesinger, Jr. and Morton White (Boston: Houghton Mifflin Company, 1963), 274–277, 283–289; Thomas Haskell, *The Emergence of Professional Social Science: The American Social Science Association and the Nineteenth-Century Crisis of Authority* (Urbana: University of Illinois Press, 1977), 192, 235–236; Dorothy Ross, "American Social Science and the Idea of Progress," in *The Authority of Experts: Studies in History and Theory,* edited by Thomas Haskell (Bloomington: Indiana University Press, 1984), 157–166; Thomas Bender, "The Erosion of Public Culture: Cities, Discourses, and Professional Disciplines," in *The Authority of Experts,* 84–101; Veysey, *Emergence of the American University,* 142–143, 173–179; Frederick Rudolph, *The American College and University: A History* (New York: Alfred A. Knopf, 1968), 245–247. See also John S. Brubacher and Willis Rudy, *Higher Education in Transition: A History of American Colleges and Universities, 1636–1968* (New York: Harper & Row, 1968), 116, 143; Roger Geiger, *To Advance Knowledge: The Growth of American Research Universities, 1900–1940* (New York: Oxford University Press, 1986), 1–3; Richard Hofstadter and C. DeWitt Hardy, *The Development and Scope of Higher Education in the United States,* 3rd ed. (New York: Columbia University Press, 1958), 31, 48, 57–58; Harold Wechsler, *The Qualified Student: A History of Selective College Admission in America* (New York: John Wiley & Sons, 1977), 8–11; Julie A. Reuben, *The Making of the Modern University: Intellectual Transformation and the Marginalization of Morality* (Chicago: The University of Chicago Press, 1996).

18. Bledstein, *Culture of Professionalism,* 33–34, 37; David K. Brown, *Degrees of Control: A Sociology of Educational Expansion and Occupational Credentialism* (New York: Teachers College Press, 1995), 10, 49–50, 61–63, 70.

19. Brown, *Degrees of Control,* 72, 137–161.

20. Zunz, *Why the American Century,* 9; Alexandra Oleson and John Voss, "Introduction," in *The Organization of Knowledge in Modern America,* xii; Geiger, *To Advance Knowledge,* 13–14; Rudolph, *American College and University,* 339–343; Bledstein, *Culture of Professionalism,* 33–39; Thelin, *History of American Higher Education,* 155; Veysey, *Emergence of the American University,* 264–266.

21. United States Bureau of Education, *Report of the Commissioner of Education for the Year 1872* (Washington, D.C.: Government Printing Office, 1873), xxxiv.

22. *Report of the Commissioner of Education, 1873*, lix, 663–682; *Report of the Commissioner of Education, 1878*, lxxx–lxxxi, xc, 515–546. Massachusetts accounted for over 80 percent of the region's city high schools in 1878.

23. *Report of the Commissioner of Education, 1873*, xlix.

24. "Schools and Scholarship," *The Nation* 19 (October 1, 1874), 215.

25. *Report of the Commissioner of Education, 1873*, 586–639.

26. S. R. Winchell, "The True Function of the High School," *Wisconsin Journal of Education* 4 (August 1874), 304.

27. *Report of the Commissioner of Education, 1873*, lix, 663–682; *Report of the Commissioner of Education, 1878*, xc, 515–546; Clifford, *"Equally in View,"* 3–4.

28. *Report of the Commissioner of Education, 1873*, 663–682; Krug, *Shaping of the American High School,* 124.

29. *Report of the Commissioner of Education, 1873*, lix, 663–682; *Report of the Commissioner of Education, 1878*, lxxx–lxxxi, xc, 515–546.

30. E. W. Coy, "Discussion," in *The Addresses and Journal of Proceedings of the National Educational Association, Session of the Year 1894* (Published by the NEA, 1895), 752. Coy was principal of Hughes High School in Cincinnati, Ohio.

31. Roger Geiger, "The Era of Multipurpose Colleges in American Higher Education, 1850–1890," in *The American College in the Nineteenth Century*, edited by Roger Geiger (Nashville: Vanderbilt University Press, 2000), 139–140; Lucas, *American Higher Education,* 131–134.

32. Geiger, "Era of Multipurpose Colleges," 135–137, 143; Roger Geiger, "Introduction," in *The American College in the Nineteenth Century*, 23–25.

33. Geiger, "Era of Multipurpose Colleges," 128–133, 141, 147–151; Roger Geiger, "The Crisis of the Old Order: The Colleges in the 1890s," in *The American College in the Nineteenth Century*, 271.

34. *Report of the Commissioner of Education, 1876*, lxx.

35. University of Michigan, *The President's Report to the Board of Regents for the Year Ending June 30, 1881* (Ann Arbor: Published by the University, 1881), 5.

36. See Veysey, *Emergence of the American University*; Brubacher and Rudy, *Higher Education in Transition;* Christopher J. Lucas, *American Higher Education: A History* (New York: St. Martin's Griffin, 1994); Reuben, *Making of the Modern University,* 5–6, 67–69, 189–191.

37. Geiger, *To Advance Knowledge,* 4–12, 16; "Introduction," 31–35; "Era of Multipurpose Colleges," 127–128; "Crisis of the Old Order," 275–276.

38. McCosh, "Upper Schools," 22–23.

39. Daniel Read, "Discussion," in *The Addresses and Journal of Proceedings of the National Educational Association, Session of the Year 1873* (The National Educational Association, 1873), 38.

40. James B. Angell, "Inaugural Address," in *Thirty-Fifth Annual Report of the Superintendent of Public Instruction of the State of Michigan* (Lansing: W.S. George & Co., 1872), 213.

41. Henry S. Frieze, "President's Annual Report," in *Thirty-fourth Annual Report of the Superintendent of Public Instruction of the State of Michigan* (Lansing: W.S. George & Co., 1870), 207.

42. Carpenter, "The Relation of the Different Educational Institutions," 86.

43. "Education," *Atlantic Monthly* 33 (May 1874), 638.

44. Hugh Hawkins, *Between Harvard and America: The Educational Leadership of Charles W. Eliot* (New York: Oxford University Press, 1972), 228.

45. Edwin C. Broome, "A Historical and Critical Discussion of College Admission Requirements" (PhD diss., Columbia University, 1902), 72–73; Reese, *Origins of the American High School*, xiii, 50, 70–72, 80.

46. *Report of the Commissioner of Education, 1887–88*, 494; *Report of the Commissioner of Education, 1890–91*, 792–795; *Report of the Commissioner of Education, 1873*, xlvii–xlviii.

47. Krug, *Shaping of the American High School*, 5–6; Reese, *Origins of the American High School*, 17, 90–94, 107.

48. Charles W. Eliot, "Present Relations of Mass. High Schools to Mass. Colleges," *Journal of Education* 21 (January 8, 1885), 19.

49. George W. Peckham, "President Bascom and the High School," *Wisconsin Journal of Education* 11 (July 1881), 302–303.

50. See Reese, *Origins of the American High School*.

51. Geo Hays and Committee, "Intermediate (or Upper) Schools," in *The Addresses and Journal of Proceedings of the National Educational Association, Session of the Year 1874* (The National Educational Association, 1874), 14.

52. Ibid., 13.

53. Andrew West, "The Relation of Secondary Education to the American University Problem," in *The Journal of Proceedings and Addresses of the National Educational Association, Session of the Year 1885* (New York: J.J. Little & Co., 1886), 201; Broome, "College Admission Requirements," 72–73; Reese, *Origins of the American High School*, 68.

54. University of Wisconsin, *Annual Report of the Board of Regents of the University of Wisconsin for the Fiscal Year Ending September 30, 1875* (Madison, WI: E.B. Bolens, State Printer, 1875), 27–28; University of Wisconsin, *Annual Report of the Board of Regents of the University of Wisconsin for the Fiscal Year Ending September 30, 1876* (Madison, WI: E. B. Bolens, State Printer, 1876), 30–31; University of Michigan, *The President's Report to the Board of Regents for the Year Ending June 30, 1872* (Ann Arbor: Published by the University, 1872), 9; University of Michigan, *The President's Report to the Board of Regents for the Year Ending June 30, 1873* (Ann Arbor: Published by the University, 1873), 7–8.

55. John Elbert Stout, *The High School: Its Function, Organization and Administration* (Boston: D. C. Heath & Co., Publishers, 1914), 66.

56. Brubacher and Rudy, *Higher Education in Transition*, 12, 245–246; Krug, *Shaping of the American High School*, 7.

57. Harvard College, *Annual Reports of the President and Treasurer of Harvard College, 1878–79* (Cambridge: University Press, 1880), 12–13.

58. Broome, "College Admission Requirements," 48–53; Brubacher and Rudy, *Higher Education in Transition*, 12, 245–246; University of Wisconsin, *Catalogue of the Officers and Students of the University of Wisconsin, for the Academic Year, 1876–7* (Madison: Democrat Printing Company, 1876), 53–54.

59. Harvard College, *Annual Reports of the President and Treasurer of Harvard College, 1872–73* (Cambridge: University Press, 1874), 19.

60. "Annual Meeting of the Classical and High-School Teachers," *New England Journal of Education* 5 (April 12, 1877), 175.

61. William C. Collar, "The Action of the Colleges upon the Schools," *Educational Review* 2 (December 1891), 426.

62. Cecil F. Bancroft, "The Service Rendered by the Secondary School," in *Sixty-Second Annual Meeting of the American Institute of Instruction, Lectures, Discussions, and Proceedings, Bethlehem, N.H., July 6–9, 1891* (Boston: American Institute of Instruction, 1891), 66–67.

63. Arthur Cohen, *The Shaping of American Higher Education: Emergence and Growth of the Contemporary System* (San Francisco: Jossey-Bass Publishers, 1998), 137–141.

64. Jurgen Herbst, *The Once and Future School: Three Hundred and Fifty Years of American Secondary Education* (New York: Routledge, 1996), 7; Broome, "College Admission Requirements," 73.

## 2  Building the University of Michigan on a High School Foundation

1. "Inspection of Bay City" (June 9, 1884), Faculty Reports, Petitions, and Resolutions. School Visits, 1883/84, Box 7, Office of the Registrar (University of Michigan), Bentley Historical Library, University of Michigan, 4–5, 6–7, 13–15.

2. Harold Wechsler, *The Qualified Student: A History of Selective College Admission in America* (New York: John Wiley & Sons, 1977), 11–12, 7–24; University of Michigan, *Annual Report of the Bureau of Co-operation with Educational Institutions*, Box 1, Bureau of School Services (University of Michigan) Records, Bentley Historical Library, University of Michigan, 6; Henry S. Frieze, "President's Annual Report," in *Thirty-Fourth Annual Report of the Superintendent of Public Instruction of the State of Michigan* (Lansing: W.S. George & Co., 1870), 209; University of Michigan, *The President's Report to the Board of Regents for the Year Ending June 30, 1871* (Ann Arbor: Published by the University, 1871), 19–22; Ewd. Olney, "The Michigan State University and the High Schools," *Wisconsin Journal of Education* 4 (July 1874), 242–243.

3. James B. Angell, "Inaugural Address, University of Michigan, 1871," chapter in *Selected Addresses* (New York: Longmans, Green, and Co., 1912), 13–14, 19–21.

4. "Secondary Education," in *Forty-Fourth Annual Report of the Superintendent of Public Instruction of the State of Michigan* (Lansing: W. S. George & Co., 1881), 345–346; "Michigan," in *Report of the Commissioner of Education for the Year 1882–83* (Washington, D.C.: Government Printing Office, 1884), 128; United States Bureau of Education, Commissioner of Education, *Report of the Commissioner of Education for the Year 1890–91* (Washington, D.C.: Government Printing Office, 1894), 792–793. As with other statistical data from the late nineteenth century, these figures have to be interpreted cautiously. These figures are based on reports submitted by school districts to the superintendent of public instruction for Michigan, but, as was often the case, not all schools submitted the necessary documents.

5. James B. Angell, *The Reminiscences of James Burrill Angell* (New York: Longmans, Green, and Co., 1912), 238–239, 255.

6. Ibid., 236–239.

7. "Report of Committee on Revision of Requirements for Admission" (June 15, 1881), Faculty Reports, Petitions, and Resolutions, Reports of Committees and Resolutions, 1880/81, Box 7, Office of the Registrar (University of Michigan), Bentley Historical Library, University of Michigan, 2.

8. University of Michigan, *The President's Report to the Board of Regents for the Year Ending June 30, 1872* (Ann Arbor: Published by the University, 1872), 10. The six Michigan schools and the number of students entering from those schools were Adrian (one), Ann Arbor (twenty-eight), Detroit (three), Flint (eight), Jackson (seven), and Kalamazoo (three); University of Michigan, *President's Report, September 30, 1890*, 11. University of Michigan, *Calendar of the University of Michigan for 1890–91* (Ann Arbor: Published by the University, 1891), 42. See University of Michigan, *Calendar of the University of Michigan* (Ann Arbor: Published by the University, 1880–1890).

9. University of Michigan, *Calendar of the University of Michigan for 1881–82* (Ann Arbor: Published by the University, 1882), 35; University of Michigan, *Calendar of the University of Michigan for 1884–85* (Ann Arbor: Published by the University, 1885), 38; "Untitled document dealing with Report of Committee on Admitting Students from out of State and from other Colleges" (May 14, 1883), Faculty Reports, Petitions, and Resolutions, General, 1882/83, Box 7, Office of the Registrar (University of Michigan), Bentley Historical Library, University of Michigan, 2–3; "Admission of Students from Schools out of the State, or from other Colleges" (January 28, 1884), Faculty Reports, Petitions, and Resolutions, General, 1883/84, Box 7, Office of the Registrar (University of Michigan), Bentley Historical Library, University of Michigan, 1–3.

10. University of Michigan, *The President's Report to the Board of Regents for the Year Ending September 30, 1891* (Ann Arbor: Published by the University, 1891), 14–15; "The University and the High School: Inspection of Schools— Extension of High School Course," *The University (of Michigan) Record* 1 (February 1892), 75.

11. University of Michigan, *Annual Report of the Bureau of Co-operation*, 9, 11; Wechsler, *Qualified Student*, 30–31.

12. "Inspection of Detroit Schools" (June 14, 1883), Faculty Reports, Petitions, and Resolutions, School Visits, 1882/83, Box 7, Office of the Registrar (University of Michigan), Bentley Historical Library, University of Michigan.

13. "Inspection of Detroit Schools" (two reports, one undated, other dated June 14, 1883), Faculty Reports, Petitions, and Resolutions, School Visits, 1882/83, Box 7, Office of the Registrar (University of Michigan), Bentley Historical Library, University of Michigan.

14. "Report of Inspecting Committee upon the High School at Big Rapids" (April 10, 1893), Faculty Reports, Petitions, and Resolutions, School Visits 1892/93, Box 8, Office of the Registrar (University of Michigan), Bentley Historical Library, University of Michigan.

15. "Inspection of Alpena" (June 9, 1884), Faculty Reports, Petitions, and Resolutions, School Visits, 1883/84, Box 7, Office of the Registrar (University of Michigan), Bentley Historical Library, University of Michigan.

16. "Lansing High School Inspection" (March 9, 1885), Faculty Reports, Petitions, and Resolutions, School Visits, 1884/85, Box 7, Office of the Registrar (University of Michigan), Bentley Historical Library, University of Michigan.

17. "Report of Inspecting Committee upon the High School at Corunna" (April 20, 1892), Faculty Reports, Petitions, and Resolutions, School Visits (folder 1), 1891/92, Box 8, Office of the Registrar (University of Michigan), Bentley Historical Library, University of Michigan; "Report of the Inspecting Committee upon the High School at Corunna" (April 20, 1893), Faculty Reports, Petitions, and Resolutions, School Visits, 1892/93, Box 8, Office of the Registrar (University of Michigan), Bentley Historical Library, University of Michigan.

18. "Report of the Inspecting Committee upon the High School at Corunna" (April 20, 1893).

19. "Records of the Committee on Diploma Schools, 1884–1893 (bound volume)," Box 1, Bureau of School Services (University of Michigan) Records, Bentley Historical Library, University of Michigan, 112–113, 194; "Report of Corunna School" (May 20, 1889), Faculty Reports, Petitions, and Resolutions, School Visits (folder 3), 1889/90, Box 8, Office of the Registrar (University of Michigan), Bentley Historical Library, University of Michigan; "Report of Inspecting Committee upon the High School at Corunna" (June 5, 1890), Faculty Reports, Petitions, and Resolutions, School Visits (folder 1), 1889/90, Box 8, Office of the Registrar (University of Michigan), Bentley Historical Library, University of Michigan; "Report of Inspecting Committee upon the High School at Corunna" (May 8, 1891), Faculty Reports, Petitions, and Resolutions, School Visits (folder 1), 1890/91, Box 8, Office of the Registrar (University of Michigan), Bentley Historical Library, University of Michigan; "Report of Inspecting Committee upon the High School at Corunna" (April 20, 1892); "Report of the Inspecting Committee upon the High School at Corunna" (April 20, 1893).

20. University of Michigan, *President's Report, June 30, 1871,* 20.

21. James B. Angell, "Relations of the University to Public Education," in *The Addresses and Journal of Proceedings of the National Educational Association, Session of the Year 1887* (Published by the NEA, 1888), 148–149.

22. See, for instance, "Holly High School Inspection" (May 9, 1884), Faculty Reports, Petitions, and Resolutions, School Visits, 1883/84, Box 7, Office of the Registrar (University of Michigan), Bentley Historical Library, University of Michigan, and "Grand Rapids High School Inspection" (May 13, 1887).

23. Angell, *Reminiscences of James Burrill Angell*, 245–246; University of Michigan, *The President's Report to the Board of Regents for the Year Ending June 30, 1879* (Ann Arbor: Published by the University, 1879), 10–11.

24. For information on the feminization of teaching and the professionalization of education, see Carl F. Kaestle, *Pillars of the Republic* (Hill and Wang, 1983); David B. Tyack, *The One Best System: A History of American Urban Education* (Cambridge: Harvard University Press, 1974); David Tyack and Elisabeth Hansot, *Managers of Virtue: Public School Leadership in American, 1820–1980* (New York: Basic Books, 1982).

25. "Eaton Rapids High School Inspection" (May 6, 1889), Faculty Reports, Petitions, and Resolutions, School Visits (folder 3), 1889/90, Box 8, Office of the Registrar (University of Michigan), Bentley Historical Library, University of Michigan; "Report of the Inspecting Committee upon the High School at Eaton Rapids, Michigan" (May 2, 1890), Faculty Reports, Petitions, and Resolutions, School Visits (folder 1), 1889/90, Box 8, Office of the Registrar (University of Michigan), Bentley Historical Library, University of Michigan.

26. "Lansing High School Inspection" (March 9, 1885).

27. Wechsler, *Qualified Student*, 17.

28. Harvard College, *Annual Reports of the President and Treasurer of Harvard College, 1873–74* (Cambridge: University Press, 1875), 11–12; Charles W. Eliot, "The Gap Between the Elementary Schools and the Colleges," in *The Addresses and Journal of Proceedings of the National Educational Association, Session of the Year 1890* (Published by the NEA, 1890), 530–531.

29. Charles W. Eliot, "Correspondence—President Eliot's Consistency," *The Nation* 26 (March 28, 1878), 212.

30. Harvard College, *Annual Reports of the President, 1873–74*, 11–12; Eliot, "Gap Between the Elementary Schools and the Colleges," 530–531.

31. Edwin C. Broome, "A Historical and Critical Discussion of College Admission Requirements" (PhD diss., Columbia University, 1902), 120–121.

32. Eliot, "Gap Between the Elementary Schools and the Colleges," 530–531.

33. For a defense of the Michigan program in light of Eliot's criticisms, see N. Dougherty, "Discussion," in *The Addresses and Journal of Proceedings of the National Educational Association, Session of the Year 1890* (Published by the NEA, 1890), 534; "Notes," *The Academy* 6 (January 1892), 581–584.

34. Charles K. Adams, "Appendix C: Reports of the Deans of Faculties, and of the Professors of the Academic Faculty," in *The President's Report to the Board of Regents for the Year Ending June 30, 1880*, University of Michigan (Ann Arbor: Published by the University, 1880), 28.

35. See, e.g., "Ann Arbor High School Inspection" (June 16, 1884), Faculty Reports, Petitions, and Resolutions, School Visits, 1883/84, Box 7, Office of the Registrar (University of Michigan), Bentley Historical Library, University of Michigan; "Inspection of Detroit Schools" (two reports, one undated, other dated June 14, 1883); "Report of Inspection of Flint High School," Faculty Reports, Petitions, and Resolutions, General, 1878/79, Box 7, Office of the Registrar (University of Michigan), Bentley Historical Library, University of Michigan; "Grand Rapids High School Inspection" (May 13, 1887).

36. Adams, "Appendix C: Reports of the Deans of Faculties, and of the Professors of the Academic Faculty," 28.

37. Ibid., 26–29.

38. Marvin Lazerson, "The College Board and American Educational History," in *A Faithful Mirror: Reflections on the College Board and Education in America,* edited by Michael C. Johanek (College Entrance Examination Board, 2001), 381; Harold Wechsler, "Eastern Standard Time: High School–College Collaboration and Admission to College, 1880–1930," in *A Faithful Mirror,* 54–55; Wechsler, *Qualified Student,* 21–23; Broome, "College Admission Requirements," 119–120; Frederick Rudolph, *The American College and University: A History* (New York: Alfred A. Knopf, 1968), 281–282; Hugh Hawkins, *Between Harvard and America: The Educational Leadership of Charles W. Eliot* (New York: Oxford University Press, 1972), 228–230.

39. "Educational Unity: The Harmonizing of Elementary, Secondary, and Collegiate Systems of Education," *Education* 3 (November 1882), 177.

40. Angell, "Relations of the University to Public Education," 149.

41. N. Dougherty, "What is the High School?" *Journal of Education* 31 (April 3, 1890), 213.

42. Joseph C. Jones, "Address to the State Teachers' Association," reprint, University of Michigan, *The President's Report to the Board of Regents for the Year Ending June 30, 1875* (Ann Arbor: Published by the University, 1875), 7.

43. Ibid.

44. Olney, "Michigan State University and the High Schools," 244.

45. Adams, "Appendix C: Reports of the Deans of Faculties, and of the Professors of the Academic Faculty," 28.

46. "Untitled document dealing with Report of Committee on Admitting Students from out of State and from other Colleges" (May 14, 1883), 1–3.

47. Wechsler, *Qualified Student,* 27–28.

48. Angell, "Relations of the University to Public Education," 149–150.

49. Ibid.

50. University of Michigan, *The President's Report to the Board of Regents for the Year Ending June 30, 1873* (Ann Arbor: Published by the University, 1873), 8.

51. University of Michigan, *The President's Report to the Board of Regents for the Year Ending September 30, 1889* (Ann Arbor: Published by the University, 1889), 12–13.

52. Ibid.

53. Charles K. Backus, "President's Report, Detroit Schools," in *Thirty-Seventh Annual Report of the Superintendent of Public Instruction of the State of Michigan* (Lansing: W.S. George & Co., 1874), 313.

54. M. J. Whitney, "Houghton," in *Thirty-Eighth Annual Report of the Superintendent of Public Instruction of the State of Michigan* (Lansing: W.S. George & Co., 1875), 288.

55. T. C. Garner, "Owosso," in *Thirty-Eighth Annual Report of the Superintendent of Public Instruction of the State of Michigan* (Lansing: W.S. George & Co., 1875), 309.

56. "Letter from C. H. McKain to James B. Angell" (February 17, 1890), Correspondence (January–February 1890), Box 3, James B. Angell Papers, 1845–1916, Bentley Historical Library, University of Michigan.

57. Angell, "Relations of the University to Public Education," 149–150.

58. Joseph C. Jones, "Pontiac," in *Thirty-Seventh Annual Report of the Superintendent of Public Instruction of the State of Michigan* (Lansing: W.S. George & Co., 1874), 349–350.

59. "Circular of the Union School at Vassar," in "Letter from E. A. Wilson to James B. Angell" (October 9, 1885), Correspondence (October–December 1885), Box 3, James B. Angell Papers, 1845–1916, Bentley Historical Library, University of Michigan.

60. "Announcement of the Vicksburg Graded School," in "Letter from C. H. McKain to James B. Angell" (February 17, 1890), Correspondence (January–February 1890), Box 3, James B. Angell Papers, 1845–1916, Bentley Historical Library, University of Michigan.

61. W. S. Perry, "Ann Arbor," in *Thirty-Eighth Annual Report of the Superintendent of Public Instruction of the State of Michigan* (Lansing: W.S. George & Co., 1875), 260.

# 3   Michigan Launches a Movement for Regional Accreditation

1. "Official Proceedings of the Wisconsin Teachers' Association—Forty-Fourth Annual Meeting," *Wisconsin Journal of Education* 27 (April 1897), 94.

2. "Admission to College by Certificate," in *Report of the Commissioner of Education for the Year 1894–95*, United States Bureau of Education (Washington, D.C.: Government Printing Office, 1896), 1172.

3. James H. Canfield, "Admission to College by Certificate—University of Nebraska," *Educational Review* 5 (March 1893), 291.

4. "Admission to College by Certificate," 1178; "Admission to College on Certificate of Secondary Schools," in *Report of the Commissioner of Education for the Year 1902*, United States Bureau of Education (Washington, D.C.: Government Printing Office, 1903), 530.

5. "Admission to College by Certificate," 1178; "Admission to College on Certificate of Secondary Schools," 530.

6. Stratton D. Brooks, "The Work of a High-School Visitor," *School Review* 9 (January 1901), 27.
7. Quoted in ibid., 27.
8. Quoted in ibid., 28.
9. Cyrus Northrop, "Admission to College by Certificate—University of Minnesota," *Educational Review* 5 (February 1893), 187–188; "Admission to College by Certificate," 1178; "Admission to College on Certificate of Secondary Schools," 530; Joseph Lindsey Henderson, *Admission to College by Certificate* (New York: Teachers College, Columbia University, 1912), 52, 60, 71; "The May Conference at the University," *Wisconsin Journal of Education* 30 (June 1900), 126–127.
10. Northrop, "Admission to College by Certificate," 188.
11. "The Indiana System of Graded Schools," in *Report of the Commissioner of Education for the Year 1886–87*, United States Bureau of Education (Washington, D.C.: Government Printing Office, 1888), 189; "Admission to College by Certificate," 1175; Elmer Ellsworth Brown, *The Making of Our Middle Schools* (New York: Longmans, Green, and Co., 1905; reprint, New York: Arno Press and *The New York Times*, 1969), 378–379; Henderson, *Admission to College by Certificate*, 54; Harold Wechsler, *The Qualified Student: A History of Selective College Admission in America* (New York: John Wiley & Sons, 1977), 41.
12. Charles W. Eliot, "The Gap Between the Elementary Schools and the Colleges," in *The Addresses and Journal of Proceedings of the National Educational Association, Session of the Year 1890* (Published by the NEA, 1890), 530–531.
13. University of Wisconsin, *Catalogue of the Officers and Students of the University of Wisconsin for the Academic Year 1876–7* (Madison, WI, 1876), 56–57.
14. Frederick Rudolph, *The American College and University: A History* (New York: Alfred A. Knopf, 1968), 281–282; Chas. W. Super, "Preparatory Departments in Connection with Colleges," *Education* 13 (September 1892), 30.
15. University of Wisconsin, *Catalogue, 1876–7*, 57.
16. University of Wisconsin, *Annual Report of the Board of Regents of the University of Wisconsin for the Fiscal Year Ending September 30, 1878* (Madison, WI: David Atwood, Printer and Stereotyper, 1878), 28.
17. University of Wisconsin, *Annual Report of the Board of Regents of the University of Wisconsin for the Fiscal Year Ending September 30, 1874* (Madison: Atwood and Culver, printers, 1874), 16; University of Wisconsin, *Annual Report, September 30, 1878*, 28.
18. "Proceedings of Wisconsin Principals' Association for 1877," *Wisconsin Journal of Education* 8 (January 1878), 46.
19. Jurgen Herbst, *The Once and Future School: Three Hundred and Fifty Years of American Secondary Education* (New York: Routledge, 1996), 83–86.
20. University of Wisconsin, *Annual Report, September 30, 1874*, 16–17.
21. University of Wisconsin, *Annual Report of the Board of Regents of the University of Wisconsin for the Fiscal Year Ending September 30, 1876* (Madison, WI: E. B. Bolens, State Printer, 1876), 31; University of Wisconsin, *Catalogue, 1876–7*, 57.

22. John Bascom, "The University and the High Schools," *Wisconsin Journal of Education* 11 (April 1881), 155; "Editorial," *Wisconsin Journal of Education* 9 (February 1879), 86.

23. "Official Proceedings of the Wisconsin Teachers' Association," 94.

24. University of Wisconsin, *Annual Report of the Board of Regents of the University of Wisconsin for the Fiscal Year Ending September 30, 1882* (Madison, WI: David Atwood, State Printer, 1882), 29.

25. University of Wisconsin, *Biennial Report of the Board of Regents of the University of Wisconsin for the Two Fiscal Years Ending September 30, 1886* (Madison, WI: Democrat Printing Company, State Printer, 1887), 35.

26. University of Wisconsin, *Catalogue of the Officers and Students of the University of Wisconsin, for the Academic Year, 1878–9* (Madison, WI: Democrat Printing Company, 1878), 46–47; University of Wisconsin, *Catalogue of the Officers and Students of the University of Wisconsin, for the Academic Year, 1884–5* (Madison, WI: Democrat Printing Company, 1884), 29–30; University of Wisconsin, *Catalogue of the Officers and Students of the University of Wisconsin, for the Academic Year, 1887–8* (Madison, WI: Democrat Printing Company, 1887), 44–46.

27. University of Michigan, *Annual Report of the Bureau of Co-operation*, 9; University of Wisconsin, *Biennial Report of the Regents of the University for the Years 1898–99 and 1899–1900* (Madison, WI: Democrat Printing Company, State Printer, 1900), 16.

28. "It Makes the Professors Laugh," *Bloomington (Wisconsin) News Item*, reprint, *Wisconsin Journal of Education* 38 (December 1906), 360.

29. University of Wisconsin, *Biennial Report, September 30, 1886*, 35.

30. University of Chicago, *The President's Report, July, 1892-July, 1902* (Chicago: The University of Chicago Press, 1903), lxvi–lxix, 211–213; Albion W. Small, "The Department of Affiliations," in *President's Report, July, 1897-July, 1898, with Summaries for 1891–7*, University of Chicago (Chicago: The University of Chicago Press, 1899), 193–196.

31. University of Chicago, *The President's Report, July, 1892-July, 1902*, lxvi–lxix, 211–213; Small, "Department of Affiliations," 193–196; "The Affiliated Work of the University," *Annual Register (University of Chicago), July, 1897-July, 1898* (Chicago: The University Press of Chicago, 1898), 131; "The Affiliated Work of the University," *Annual Register (University of Chicago), July, 1898-July, 1899* (Chicago: The University Press of Chicago, 1899), 132; William R. Harper, "Entrance Requirements—The Chicago System," in *Journal of Proceedings and Addresses of the Thirty-Fifth Annual Meeting*, National Education Association (Published by the NEA, 1896), 631–632.

32. University of Chicago, *The President's Report, July, 1892-July, 1902*, lxvi–lxix; Frank J. Miller, "The University Affiliations," in *The President's Report, July, 1892-July, 1902*, University of Chicago (Chicago: The University of Chicago Press, 1903), 211–213; Small, "Department of Affiliations," 193–196.

33. Small, "Department of Affiliations," 197.

34. University of Chicago, *The President's Report, July, 1898-July, 1899* (Chicago: The University of Chicago Press, 1900), xi.

35. W. D. MacClintock, "University College," in *The President's Report, July, 1892-July, 1902*, University of Chicago (Chicago: The University of Chicago Press, 1903), 146, 152–155; Edmund J. James, "The College for Teachers," in *The President's Report, July, 1898-July, 1899*, University of Chicago (Chicago: The University of Chicago Press, 1900), 79–80.

36. Edwin C. Broome, "A Historical and Critical Discussion of College Admission Requirements" (PhD diss., Columbia University, 1902), 121; Brown, *Making of Our Middle Schools*, 374–375; Leon J. Richardson, "The University of California and the Accrediting of Secondary Schools," *School Review* 10 (October 1902), 615–616.

37. Martin Kellogg, "Admission to College by Certificate—Accredited Schools in California," *Educational Review* 5 (April 1893), 385.

38. Ibid., 386; William Carey Jones, "The Prospects for a Federal Educational Union," in *Journal of Proceedings and Addresses, Session of the Year 1895*, National Education Association (Published by the NEA, 1895), 596; "Admission to College on Certificate of Secondary Schools," 529.

39. Eliot, "Gap Between the Elementary Schools and the Colleges," 530–531.

40. See, for instance, "Report of the Committee on Accredited Schools, January 27, 1896," Box 1, University of Wisconsin Faculty Documents, 1889–1967, Series 5/2/2/2, University of Wisconsin–Madison Archives, Memorial Library, University of Wisconsin–Madison; "Report of the Committee on Accredited Schools, April 20, 1896," Box 1, University of Wisconsin Faculty Documents, 1889–1967, Series 5/2/2/2, University of Wisconsin–Madison Archives, Memorial Library, University of Wisconsin–Madison; "Report of the Committee on Accredited Schools, March 15, 1897," Box 1, University of Wisconsin Faculty Documents, 1889–1967, Series 5/2/2/2, University of Wisconsin–Madison Archives, Memorial Library, University of Wisconsin–Madison.

41. Northrop, "Admission to College by Certificate," 187; "Admission to College by Certificate," 1178; "Admission to College on Certificate of Secondary Schools," 530.

42. "Admission to College by Certificate," 1182.

43. Ibid., 1171, 1184–1187.

44. Donald C. Agnew, *Seventy-Five Years of Educational Leadership* (Atlanta: Southern Association of Colleges and Schools, 1970), 5; Horace Mann Bond, *The Education of the Negro in the American Social Order* (New York: Octagan Books, 1969), 92–93; James Anderson, *The Education of Blacks in the South, 1860–1935* (Chapel Hill: The University of North Carolina Press, 1988), 81–86, 110–115, 137–153, 186–191, 238–249.

45. E. S. Joynes, quoted in Charles Forster Smith, "Southern Colleges and Schools," *Atlantic Monthly* 56 (December 1885), 740.

46. Edwin Mims, *Chancellor Kirkland of Vanderbilt* (Nashville: Vanderbilt University Press, 1940), 133–134.

47. "Admission to College by Certificate," 1184.

48. Paul H. Saunders, "The Outlook of the Public High School in the South," in *Proceedings of the Eighth Annual Meeting*, Association of Colleges and

Preparatory Schools of the Southern States (Printed by the University of Chicago Press, n.d.), 92.

49. "Admission to College by Certificate," 1173; Saunders, "Outlook of the Public High School in the South," 92.

50. "Admission to College by Certificate," 1184.

51. Ibid., 1178; "Admission to College on Certificate of Secondary Schools," 530.

52. R. Bingham, quoted in Charles Forster Smith, "Southern Colleges and Schools," *Atlantic Monthly* 56 (December 1885), 742.

53. Addison Hogue, "Should the Southern College Association in Its By-laws Forbid Preparatory Departments and Require Entrance Examinations for Admission to College?," in *Proceedings of the Sixth Annual Meeting*, Association of Colleges and Preparatory Schools of the Southern States (Printed by the University of Chicago Press, n.d.), 17–18.

54. Edward Mayes, "History of Education in Mississippi," in *Circular of Information, no. 2, 1899*, United States Bureau of Education (Washington, D.C.: Government Printing Office, 1899), 174–175.

55. Saunders, "Outlook of the Public High School in the South," 82.

56. T. W. Jordan, "Admission to College on Certificate," in *Proceedings of the Fifth Annual Meeting*, Association of Colleges and Preparatory Schools of the Southern States (n.p., n.d.), 83–84; Saunders, "Outlook of the Public High School in the South," 92.

57. Jordan, "Admission to College on Certificate," 84.

58. C. B. Wallace, "Admission to College on Certificate," in *Proceedings of the Fifth Annual Meeting*, Association of Colleges and Preparatory Schools of the Southern States (n.p., n.d.), 87.

59. Josiah H. Shinn, "History of Education in Arkansas," in *Circular of Information, no. 1, 1900*, United States Bureau of Education (Washington, D.C.: Government Printing Office, 1900), 95; "Admission to College on Certificate of Secondary Schools," 529.

60. Wallace, "Admission to College on Certificate," 87–88.

61. R. W. Jones, "Our Proposed New Requirements for Admission to College," in *Proceedings of the Sixth Annual Meeting*, Association of Colleges and Preparatory Schools of the Southern States (Printed by the University of Chicago Press, n.d.), 11–12.

62. Bond, *Education of the Negro in the American Social Order*, 88, 92–96, 101–103, 108–109; J. Morgan Kousser, "Progressivism—For Middle-Class Whites Only: North Carolina Education, 1880–1910," *The Journal of Southern History* 46 (May 1980), 173–175, 181. For information on the place of African Americans in the South following the Civil War, see, e.g., George Fredrickson, *The Black Image in the White Mind: The Debate on Afro-American Character and Destiny, 1817–1914* (New York: Harper & Row, Publishers, 1971); C. Vann Woodward, *Origins of the New South, 1877–1913* (Baton Rouge: Louisiana State University Press, 1971); Leon Litwack, *Trouble in Mind: Black Southerners in the Age of Jim Crow* (New York: Alfred A. Knopf, 1998); Paul Gaston, *The New South Creed: A Study in Southern Mythmaking* (New York: Knopf, 1970); Edward Ayers, *The Promise of the New South: Life after*

*Reconstruction* (New York: Oxford University Press, 1992); George Fredrickson, *The Arrogance of Race: Historical Perspectives on Slavery, Racism, and Social Inequality* (Middletown, CT: Wesleyan University Press, 1988); Joel Williamson, *The Crucible of Race: Black–White Relations in the American South since Emancipation* (New York: Oxford University Press, 1984); Howard Rabinowitz, *Race Relations in the Urban South, 1865–1890* (New York: Oxford University Press, 1978).

63. Anderson, *Education of Blacks in the South*, 148–152, 187–191.

64. John S. Brubacher and Willis Rudy, *Higher Education in Transition: A History of American Colleges and Universities, 1636–1968* (New York: Harper & Row, 1968), 75; John R. Thelin, *A History of American Higher Education* (Baltimore: The Johns Hopkins University Press, 2004), 186; Anderson, *Education of Blacks in the South*, 249.

65. Edwin Whitfield Fay, "The History of Education in Louisiana," in *Circular of Information, no. 1, 1898,* United States Bureau of Education (Washington, D.C.: Government Printing Office, 1898), 149–151.

66. "Admission to College by Certificate," 1182.

67. Leland Stanford Cozart, *A History of the Association of Colleges and Secondary Schools, 1934–1965* (Charlotte, NC: Heritage Printers, Inc., 1967), 1–2.

68. President's comments quoted in "Admission into College by Certificates," *Journal of Education* 26 (September 22, 1887), 167.

69. O. M. Fernald, "Admission to College by Certificate—Williams College," *Educational Review* 5 (March 1893), 292–295.

70. "Admission into College by Certificates," *Journal of Education* 26 (September 22, 1887), 167; Merrill Gates, "Admission to College by Certificate—Amherst College," *Educational Review* 5 (February 1893), 189–191; John K. Lord, "Admission to College by Certificate," *New England Journal of Education* 10 (September 11, 1879), 129; Lucy M. Salmon, "Unity of Standard for College Entrance Examination," *The Academy* 3 (May 1888), 222–231; Lucy M. Salmon, "Different Methods of Admission to College," *Educational Review* 6 (October 1893), 231; William C. Collar, "The Action of the Colleges upon the Schools," *Educational Review* 2 (December 1891), 433–434; "Admission to College by Certificate," 1172–1173, 1184.

71. Eliot, "Gap Between the Elementary Schools and the Colleges," 530.

72. Harvard College, *Annual Reports of the President and Treasurer of Harvard College, 1891–92* (Cambridge: Published by the University, 1893), 14–15; Paul Hanus, "University Inspection of Secondary Schools and the Schools Examination Board of Harvard University," *School Review* 2 (May 1894), 260–264.

73. Isaac Thomas, "Some Ways in Which College May Help Secondary Schools," *The Academy* 7 (April 1892), 151.

74. John Tetlow, "Admission to College by Certificate—Girls' Latin School, Boston," *Educational Review* 5 (April 1893), 389–390.

75. Charles W. Eliot, "The New England Association of Colleges and Preparatory Schools," *School and College* 1 (November 1892), 563.

76. Charles F. Dunbar, "Report of the Dean of the College Faculty," in *Annual Reports of the President and Treasurer of Harvard College, 1876–77* (Cambridge: University Press, 1878), 59; Charles F. Dunbar, "Report of the Dean of the College Faculty," in *Annual Reports of the President and Treasurer of Harvard College, 1877–78* (Cambridge: University Press, 1879), 66; Harvard College, *Annual Reports of the President and Treasurer of Harvard College, 1878–79* (Cambridge: University Press, 1880), 10; Charles F. Dunbar, "Report of the Dean of the College Faculty," in *Annual Reports of the President and Treasurer of Harvard College, 1878–79* (Cambridge: University Press, 1880), 67.

77. Collar, "Action of the Colleges upon the Schools," 435.

78. William C. Collar, "The New England Association of Colleges and Preparatory Schools," *School and College* 1 (November 1892), 560.

79. William F. Bradbury, "The New England Association of Colleges and Preparatory Schools," *School and College* 1 (November 1892), 560.

80. Horace M. Willard, "Is The System of Admission to Colleges on School Certificates Advantageous to Schools?," in *Addresses and Proceedings of the Preliminary Meeting, Oct. 16 and 17, 1885, First Annual Meeting, Oct. 16, 1886, and First Special Meeting, Jan. 7 and 8, 1887*, New England Association of Colleges and Preparatory Schools (Boston, Mass: Published by the Association, 1892), 39.

81. Alice Freeman, "Discussion," in *Addresses and Proceedings of the Preliminary Meeting, Oct. 16 and 17, 1885, First Annual Meeting, Oct. 16, 1886, and First Special Meeting, Jan. 7 and 8, 1887*, New England Association of Colleges and Preparatory Schools (Boston, Mass: Published by the Association, 1892), 40.

82. Lord, "Admission to College by Certificate," 128–129.

83. L. Clark Seelye, "The New England Association of Colleges and Preparatory Schools," *School and College* 1 (November 1892), 562.

84. Andrea Walton, "Cultivating a Place for Selective All-Female Education in a Coeducational World: Women Educators and Professional voluntary Associations, 1880–1926," in *A Faithful Mirror: Reflections on the College Board and Education in America*, edited by Michael C. Johanek (College Entrance Examination Board, 2001), 155; William J. Reese, *The Origins of the American High School* (New Haven: Yale University Press, 1995), 222–225.

85. James Taylor, "Discussion," in *Report of the Fifteenth Annual Meeting of the New England Association of Colleges and Preparatory Schools, Held at Boston, Mass., Oct. 12 and 13, 1900*, New England Association of Colleges and Preparatory Schools, (Chicago: Reprinted from *School Review*, 1900), 64.

86. Francis A. Waterhouse, "The New England Association of Colleges and Preparatory Schools," *School and College* 1 (November 1892), 562.

87. Commission of Colleges in New England on Admission Examinations, *Fifteenth Annual Report, 1900–1901* (Providence: Snow & Farnham, Printers, 1901), 10.

88. Lord, "Admission to College by Certificate," 128.

89. "Editorial," Educational Review 6 (October 1893), 308–309.

90. Cecil F. Bancroft, "Is Any Greater Degree of Uniformity in Requisitions for Admission to College Practicable?," in *Addresses and Proceedings of the Preliminary Meeting, Oct. 16 and 17, 1885, First Annual Meeting, Oct. 16, 1886, and First Special Meeting, Jan. 7 and 8, 1887*, New England Association of Colleges and Preparatory Schools (Boston, Mass: Published by the Association, 1892), 10.

91. Waterhouse, "Methods of Determining the Qualifications of Candidates for Admission to College," *School and College* 1 (November 1892), 533–534.

92. New England Association of Colleges and Preparatory Schools, *Addresses and Proceedings of the Preliminary Meeting, Oct. 16 and 17, 1885, First Annual Meeting, Oct. 16, 1886, and First Special Meeting, Jan. 7 and 8, 1887* (Boston, Mass: Published by the Association, 1892), 4.

93. Bancroft, "Is Any Greater Degree of Uniformity in Requisitions for Admission to College Practicable?," 11.

94. Ibid.

95. New England Association of Colleges and Preparatory Schools, *Addresses and Proceedings of the Preliminary Meeting, Oct. 16 and 17, 1885, First Annual Meeting, Oct. 16, 1886, and First Special Meeting, Jan. 7 and 8, 1887*, 34.

96. Ibid., 36–37; "College Admission Examinations: Report of the Commission of Colleges in New England," *The Academy* 2 (March 1887), 69–77.

97. New England Association of Colleges and Preparatory Schools, "Official Report of the Third Annual Meeting of the New England Association of Colleges and Preparatory Schools," *The Academy* 3 (November 1888), 504–506; Commission of Colleges in New England on Admission Examinations, *Second Annual Report, 1887–1888* (Providence: Snow & Farnham, Printers, 1889), 7–16; New England Association of Colleges and Preparatory Schools, *Addresses and Proceedings of the Fourth Annual Meeting Held at Boston, October 11th and 12th, 1889* (Syracuse, NY: Reprinted from The Academy, 1889), 47.

98. Commission of Colleges in New England on Admission Examinations, *Eighth Annual Report, 1893–1894* (Providence: Snow & Farnham, Printers, 1894), 12–27.

99. John Tetlow, "High Schools," *Journal of Education* 44 (August 13, 1896), 114.

100. Commission of Colleges in New England on Admission Examinations, *Ninth Annual Report, 1894–1895* (Providence: Snow & Farnham, Printers, 1895), 7–8, 13.

101. Commission of Colleges in New England on Admission Examinations, *Tenth Annual Report, 1895–1896* (Providence: Snow & Farnham, Printers, 1896), 11.

102. Commission of Colleges in New England on Admission Examinations, *Eleventh Annual Report, 1896–1897* (Providence: Snow & Farnham, Printers, 1897), 12–13, 15.

103. New England Association of Colleges and Preparatory Schools, *Official Report of the Fifth Annual Meeting of the N. E. Association of Colleges and Preparatory Schools, Held at Boston, Oct. 17 and 18, 1890* (Boston: Reprinted from *The Academy*, 1890), 45.

104. Commission of Colleges in New England on Admission Examinations, *Fifteenth Annual Report*, 10.

105. Robert Keep, "Under What Conditions Might Admission to College by Certificate Be Permitted?" in *Addresses and Proceedings of the Preliminary Meeting, Oct. 16 and 17, 1885, First Annual Meeting, Oct. 16, 1886, and First Special Meeting, Jan. 7 and 8, 1887*, New England Association of Colleges and Preparatory Schools (Boston, Mass: Published by the Association, 1892), 28–30.

106. James B. Angell, "Relations of the University to Public Education," in *The Addresses and Journal of Proceedings of the National Educational Association, Session of the Year 1887* (Published by the NEA, 1888), 150.

107. Commission of Colleges in New England on Admission Examinations, *Fifteenth Annual Report*, 7–16; Commission of Colleges in New England on Admission Examinations, *Sixteenth Annual Report, 1901–1902* (Providence: Snow & Farnham, Printers, 1902), 7–10; "Admission to College on Certificate of Secondary Schools," 527–533.

# 4   The Secondary Schools' Challenge to Higher Education and the Dominance of the Modern Subjects

1. "Letter to William Rainey Harper from J. A. Eberly, June 30, 1898," Presidents Papers, 1889–1925, Box 9, Folder 5, Special Collections Research Center, University of Chicago Library.

2. University of Chicago, "The First Annual Report of the University of Chicago (1892)," Special Collections Research Center, Reference, University of Chicago Library.

3. Charles W. Eliot, "Discussion," in *The Addresses and Journal of Proceedings of the National Educational Association, Session of the Year 1873* (The National Educational Association, 1873), 44.

4. Charles W. Eliot, quoted in "Preparatory Schools," *Wisconsin Journal of Education* 6 (March 1876), 101; "Annual Meeting of the Classical and High-School Teachers," 175.

5. University of Wisconsin, *Annual Report of the Board of Regents of the University of Wisconsin for the Fiscal Year Ending September 30, 1880* (Madison, WI: David Atwood, State Printer, 1880), 30–31.

6. George W. Peckham, "President Bascom and the High School," *Wisconsin Journal of Education* 11 (July 1881), 303–304.

7. Harold Wechsler, *The Qualified Student: A History of Selective College Admission in America* (New York: John Wiley & Sons, 1977), 17; for the emergence of public high schools, see William J. Reese, *The Origins of the American High School* (New Haven: Yale University Press, 1995).

8. Reese, *Origins of the American High School*, 90–118.

9. W. S. Perry, "Ann Arbor," in *Thirty-Eighth Annual Report of the Superintendent of Public Instruction of the State of Michigan* (Lansing: W.S. George & Co., 1875), 259.

10. S. S. Babcock, "The Independent Work of the High School," in *Thirty-Eighth Annual Report of the Superintendent of Public Instruction of the State of Michigan* (Lansing: W.S. George & Co., 1875), 373.

11. I. L. Stone, "Battle Creek," in *Thirty-Ninth Annual Report of the Superintendent of Public Instruction of the State of Michigan* (Lansing: W.S. George & Co., 1876), 319.

12. H. J. Taylor, "The Function of the High School," in *Wisconsin Journal of Education* 13 (December 1883), 485–486.

13. "Editorial," *Wisconsin Journal of Education* 15 (November 1885), 485.

14. Bernard Bigsby, "Port Huron," in *Thirty-Ninth Annual Report of the Superintendent of Public Instruction of the State of Michigan* (Lansing: W.S. George & Co., 1876), 406.

15. J. Fairbanks, "The High School and the College," *Journal of Education* 21 (March 5, 1885), 147.

16. Charles W. Eliot, "Present Relations of Mass. High Schools to Mass. Colleges," *Journal of Education* 21 (January 8, 1885), 19.

17. Harvard College, *Annual Reports of the President and Treasurer of Harvard College, 1884–85* (Cambridge: University Press, 1886), 50, 189–193; Hugh Hawkins, *Between Harvard and America: The Educational Leadership of Charles W. Eliot* (New York: Oxford University Press, 1972), 228–231.

18. Eliot, "Present Relations," 19–20.

19. Ibid., 20.

20. Hawkins, *Between Harvard and America*, 234–235; Eliot, "Present Relations," 20.

21. Eliot, "Present Relations," 19.

22. Ibid., 20.

23. Ibid.; David L. Angus and Jeffrey E. Mirel, "Presidents, Professors, and Lay Boards of Education: The Struggle for Influence over the American High School, 1860–1910," in *A Faithful Mirror: Reflections on the College Board and Education in America*, edited by Michael C. Johanek (College Entrance Examination Board, 2001), 15.

24. The University of Chicago, *President's Report, July, 1897–July, 1898, with Summaries for 1891–7* (Chicago: The University of Chicago Press, 1899), 80.

25. "Report of the Seventh Semi-Annual Educational Conference," *University (of Chicago) Record* 1 (June 12, 1896), 195.

26. Harvard College, *Annual Reports of the President and Treasurer of Harvard College, 1884–85* (Cambridge: University Press, 1886), 49.

27. Edwin C. Broome, "A Historical and Critical Discussion of College Admission Requirements" (PhD diss., Columbia University, 1902), 46–54, 61; Richard

Hofstadter and C. DeWitt Hardy, *The Development and Scope of Higher Education in the United States,* 3rd ed. (New York: Columbia University Press, 1958), 10–11.

28. Broome, "College Admission Requirements," 77, 82; Elmer Ellsworth Brown, *The Making of Our Middle Schools* (New York: Longmans, Green, and Co., 1905; reprint, New York: Arno Press and *The New York Times,* 1969), 372–373; John S. Brubacher and Willis Rudy, *Higher Education in Transition: A History of American Colleges and Universities, 1636–1968* (New York: Harper & Row, 1968), 110; Harvard College, *Annual Reports of the President and Treasurer of Harvard College, 1892–93* (Cambridge: Published by the University, 1894), 10–11; Merle Curti and Vernon Carstensen, *The University of Wisconsin: A History, 1848–1925,* vol. 2 (Madison: University of Wisconsin Press, 1949), 315.

29. University of Michigan, *The President's Report to the Board of Regents for the Year Ending June 30, 1875* (Ann Arbor: Published by the University, 1875), 9.

30. University of Michigan, *The President's Report to the Board of Regents for the Year Ending June 30, 1878* (Ann Arbor: Published by the University, 1878), 7–8; University of Michigan, *The President's Report to the Board of Regents for the Year Ending June 30, 1879* (Ann Arbor: Published by the University, 1879), 6; University of Michigan, *Calendar of the University of Michigan for 1880–81* (Ann Arbor: Published by the University, 1881), 28–32.

31. Frederick Jackson Turner, "The Significance of History," in *The Early Writings of Frederick Jackson Turner,* edited by Everett Edwards (Madison: University of Wisconsin Press, 1938); Broome, "College Admission Requirements," 68, 83–85.

32. Brown, *Making of Our Middle Schools,* 372–373; Broome, "College Admission Requirements," 83–85; Harold Wechsler, "Eastern Standard Time: High School-College Collaboration and Admission to College, 1880–1930," in *A Faithful Mirror,* 45.

33. Broome, "College Admission Requirements," 78–82.

34. R. S. Keyser, "College Preparation and the Public Schools," *The Academy* 2 (April 1887), 113.

35. "What the Secondary Schools Ask of the Colleges," *The Academy* 2 (April 1887), 134.

36. "Proposal for a Report of the Committee on Revision of Requirements for Admission" (February 27, 1888), Faculty Reports, Petitions, and Resolutions. Faculty Meetings (folder 1), 1888/89, Box 8, Office of the Registrar (University of Michigan), Bentley Historical Library, University of Michigan, 2–3.

37. "Proposal for a Report of the Committee on Revision of Requirements for Admission" (February 27, 1888), 2–3.

38. "Letter to the Faculty of the University of Michigan from Ann Arbor High School Teachers" (not dated, perhaps March 1888), Faculty Reports, Petitions, and Resolutions. Faculty Meetings (folder 1), 1888/89, Box 8, Office of the Registrar (University of Michigan), Bentley Historical Library, University of Michigan.

39. "Announcement to Preparatory Schools" (December 1, 1888), Faculty Reports, Petitions, and Resolutions. Faculty Meetings (folder 1), 1888/89, Box 8, Office of the Registrar (University of Michigan), Bentley Historical Library, University of Michigan, 2; University of Michigan, *The President's Report to the Board of Regents for the Year Ending September 30, 1889* (Ann Arbor: Published by the University, 1889), 11–13.

40. "Letter from T. C. Chamberlin to Principals of Accredited Schools" (April 7, 1890), Box 1, University of Wisconsin Faculty Documents, 1889–1967, Series 5/2/2/2, University of Wisconsin–Madison Archives, Memorial Library, University of Wisconsin–Madison.

41. University of Wisconsin, *Biennial Report of the Board of Regents of the University of Wisconsin for the Two Fiscal Years Ending September 30, 1892* (Madison, WI: Democrat Printing Company, State Printers, 1893), 40–41.

42. Hawkins, *Between Harvard and America*, 171–174.

43. Eliot, "Present Relations," 20.

44. Hawkins, *Between Harvard and America*, 171–174.

45. Keyser, "College Preparation and the Public Schools," 114–115.

46. Charles W. Eliot, "Admission Requirements," in *Report of the Commissioner of Education for the Year 1885–86*, United States Bureau of Education (Washington, D.C.: Government Printing Office, 1887), 471; Hawkins, *Between Harvard and America*, 175–176.

47. United States Bureau of Education, *Report of the Commissioner of Education for the Year 1888–89* (Washington, D.C.: Government Printing Office, 1891), liv.

48. Cecil F. Bancroft, "The Service Rendered by the Secondary School," in *Sixty-Second Annual Meeting of the American Institute of Instruction, Lectures, Discussions, and Proceedings, Bethlehem, N.H., July 6–9, 1891* (Boston: American Institute of Instruction, 1891), 69.

49. Ibid., 70–72.

50. Ibid., 73.

51. Isaac Thomas, "Some Ways in Which College May Help Secondary Schools," *The Academy* 7 (April 1892), 146.

52. E. S. Hawes, "The Place of Work Preparatory for the College in Relation to Other Work in the Secondary Schools," *The Academy* 7 (April 1892), 164.

# 5  Charles W. Eliot and the Early Campaign for a National Educational System

1. James M. Greenwood, "Discussion," in *The Addresses and Journal of Proceedings of the National Educational Association, Session of the Year 1894* (Published by the NEA, 1895), 454.

2. "Letter from Charles W. Eliot to James B. Angell (April 11, 1893)," Correspondence (April 1893—Folder 132), Box 4, James B. Angell Papers,

Bentley Historical Library, University of Michigan. The NEA provided $2,500 and private contributions accounted for another $1,899. Eliot wrote Angell that of this sum, only $650 was left over to pay for printing costs.

3. Greenwood, "Discussion," 454.

4. Charles W. Eliot, "The Report of the Committee of Ten," *Educational Review* 7 (February 1894), 105; Nicholas Murray Butler, "The Reform of Secondary Education in the United States," *The Atlantic Monthly* 73 (March 1894), 373.

5. "Report of Committee on Secondary Education, Uniformity in Requirements for Admission to College," in *The Addresses and Journal of Proceedings of the National Educational Association, Session of the Year 1891* (Published by the NEA, 1891), 310.

6. Ibid., 315.

7. William T. Harris, "The Committee of Ten on Secondary Schools," *Educational Review* 7 (January 1894), 1–2.

8. National Education Association, *Report of the Committee of Ten on Secondary School Studies, with the Report of the Conferences Arranged by the Committee* (New York: American Book Company, 1894), 3–4; "Report of Committee on Secondary Education, Uniformity in Requirements for Admission to College," 279, 316, 695; "Report of the Secretary of the National Council of Education," in *The Addresses and Journal of Proceedings of the National Educational Association, Session of the Year 1892* (Published by the NEA, 1892), 754; "Minutes of the Board of Directors," in *The Addresses and Journal of Proceedings of the National Educational Association, Session of the Year 1892* (Published by the NEA, 1892), 31.

9. "Minutes of the Board of Directors," 31; "The New England Association of Colleges and Preparatory Schools," *School Review* 1 (December 1893), 603; Theodore R. Sizer, *Secondary Schools at the Turn of the Century* (New Haven: Yale University Press, 1964), 107.

10. "Letter from Charles W. Eliot to James B. Angell (July 19, 1892)," Correspondence (July–September 1892—Folder 128), Box 4, James B. Angell Papers, Bentley Historical Library, University of Michigan.

11. National Education Association, *Report of the Committee of Ten on Secondary School Studies,* 11; "Report of the Secretary of the National Council of Education," 754.

12. Harvard College, *Annual Reports of the President and Treasurer of Harvard College, 1894–95* (Cambridge: Published by the University, 1896), 9–10.

13. Ray Greene Huling, "The Reports on Secondary School Studies," *School Review* 2 (May 1894), 278–279. As principal of the New Bedford, Massachusetts, High School, he was a member of the conference on history, civil government, and political economy.

14. "Letter from Charles W. Eliot," High School Material (Folder 510), Box 11, James B. Angell Papers, Bentley Historical Library, University of Michigan.

15. National Education Association, *Report of the Committee of Ten on Secondary School Studies,* 6.

16. National Education Association, *Report of the Committee of Ten on Secondary School Studies,* 3–12.

17. "On Secondary Education: National Educational Society's Committee Finishes Its Work," *New York Times* (November 13, 1893), 5.

18. National Education Association, *Report of the Committee of Ten on Secondary School Studies*, 46–47.

19. Ibid., 45–48.

20. Harvard College, *Annual Reports, 1894–95*, 76–77.

21. Harvard College, *Annual Reports of the President and Treasurer of Harvard College, 1897–98* (Cambridge: Published by the University, 1899), 101; Harvard College, *Annual Reports of the President and Treasurer of Harvard College, 1898–99* (Cambridge: Published by the University, 1900), 7–8.

22. Hugh Hawkins, *Between Harvard and America: The Educational Leadership of Charles W. Eliot* (New York: Oxford University Press, 1972), 175–176.

23. Harvard College, *Reports of the President and Treasurer of Harvard College, 1896–97* (Cambridge: Published by the University, 1898), 120–122; Harvard College, *Annual Reports, 1897–98*, 24–25.

24. Jurgen Herbst, *The Once and Future School: Three Hundred and Fifty Years of American Secondary Education* (New York: Routledge, 1996), 112–113; Hawkins, *Between Harvard and America*, 232–233; Krug, *Shaping of the American High School*, 89–90; Sizer, *Secondary Schools at the Turn of the Century*, 185.

25. See, e.g., David L. Angus and Jeffrey E. Mirel, "Presidents, Professors, and Lay Boards of Education: The Struggle for Influence over the American High School, 1860–1910," in *A Faithful Mirror: Reflections on the College Board and Education in America*, edited by Michael C. Johanek (College Entrance Examination Board, 2001); Elmer Ellsworth Brown, *The Making of Our Middle Schools* (New York: Longmans, Green, and Co., 1905; reprint, New York: Arno Press and The New York Times, 1969), 384; Edward A. Krug, *The Shaping of the American High School, 1880–1920* (Madison: The University of Wisconsin Press, 1969), 88–90; Sizer, *Secondary Schools at the Turn of the Century*, 184–192; Edwin G. Dexter, "Ten Years' Influence of the Report of the Committee of Ten," *School Review* 14 (April 1906), 269.

26. "Report of the Committee on Accredited Schools, in re President Eliot's Letter, June 18, 1894," Box 1, University of Wisconsin Faculty Documents, 1889–1967, Series 5/2/2/2, University of Wisconsin–Madison Archives, Memorial Library, University of Wisconsin–Madison. See also Charles W. Eliot and John Tetlow, "Letter to Faculties of Professional, Scientific, and Technological Schools, May 1, 1894," Box 1, University of Wisconsin Faculty Documents, 1889–1967, Series 5/2/2/2, University of Wisconsin–Madison Archives, Memorial Library, University of Wisconsin–Madison.

27. "Wisconsin Courses and the Report of the Committee of Ten," *Wisconsin Journal of Education* 24 (March 1894), 51–52.

28. Angus and Mirel, "Presidents, Professors, and Lay Boards of Education," 25; Brown, *Making of Our Middle Schools*, 384; Krug, *Shaping of the American High School*, 88–90; Sizer, *Secondary Schools at the Turn of the Century*, 184–192. Sizer is more willing than the other authors to suggest that the report had some effect on school curricula, but even he admits that this influence was

slight and less significant than its influence on general educational policy and thinking, especially as such thinking revolved around electives and equivalence of subjects.

29. Dexter, "Ten Years' Influence of the Report of the Committee of Ten," 269.
30. "The New England Association of Colleges and Preparatory Schools," *School Review* 2 (December 1894), 621, 661; "The New England Association of Colleges and Preparatory Schools, Second Special Meeting," *School Review* 3 (March 1895), 181–184; "New England Association of Colleges and Preparatory Schools," Letter Press Volume, 1889–1898, Charles W. Eliot Papers, Harvard University Archives, Harvard University, 91–92.
31. National Education Association, *Report of the Committee on College Entrance Requirements, July, 1899* (Published by the Association, 1899), 5.
32. Ibid., 6, 10–11.
33. G. Stanley Hall, "Discussion," in *The Addresses and Journal of Proceedings of the National Educational Association, Session of the Year 1894* (Published by the NEA, 1895), 663.
34. "Letter from James B. Angell to B. L. D'Ooge (March 26, 1894)," reprint, *The Inlander: Supplement* 4 (June 1894), 56.
35. "Proceedings of the Michigan Schoolmasters' Club," *The Inlander: Supplement* 4 (June 1894), 47.
36. James C. Mackenzie, "The Report of the Committee of Ten," *School Review* 2 (March 1894), 148.
37. Charles W. Eliot, "Secondary School Programs, and the Conferences of December, 1892," in "The New England Association of Colleges and Preparatory Schools," *School Review* 1 (December 1893), 613.
38. "Letter to Seth Low from Charles W. Eliot, October 26, 1895," Letter Press Volume, 1889–1898, Charles W. Eliot Papers, Harvard University Archives, Harvard University, 108.
39. National Education Association, *Report of the Committee of Ten on Secondary School Studies*, 52.
40. Ibid., 51–53.
41. Ibid., 51–52.
42. Ibid., 17.
43. Eliot, "Report of the Committee of Ten," 109.
44. Henry King, "Discussion," in *The Addresses and Journal of Proceedings of the National Educational Association, Session of the Year 1894* (Published by the NEA, 1895), 666.
45. H. S. Tarbell, "Discussion," in *The Addresses and Journal of Proceedings of the National Educational Association, Session of the Year 1894* (Published by the NEA, 1895), 665.
46. C. M. Woodward, "Discussion," in *The Addresses and Journal of Proceedings of the National Educational Association, Session of the Year 1894* (Published by the NEA, 1895), 661; James C. Mackenzie, "Discussion," in *The Addresses and Journal of Proceedings of the National Educational Association, Session of the Year 1894* (Published by the NEA, 1895), 662–663; Angus and Mirel, "Presidents, Professors, and Lay Boards of Education," 22.

47. Oscar D. Robinson, "The Work of the Committee of Ten," *School Review* 2 (June 1894), 366–372.

48. National Education Association, *Report of the Committee of Ten on Secondary School Studies*, 52; United States Bureau of Education, *Report of the Commissioner of Education for the Year 1890–91* (Washington, D.C.: Government Printing Office, 1894), 792, 794, 796.

49. Eliot, "Report of the Committee of Ten," 106.

50. National Education Association, *Report of the Committee of Ten on Secondary School Studies*, 40–41.

51. Herbert Kliebard, *The Struggle for the American Curriculum, 1893–1958* (New York: Routledge, 1995), 9.

52. "Letter to Seth Low from Charles W. Eliot, October 26, 1895," 108.

53. National Education Association, *Report of the Committee of Ten on Secondary School Studies*, 43.

54. David B. Tyack, *The One Best System: A History of American Urban Education* (Cambridge: Harvard University Press, 1974), 126–127.

55. Butler, "Reform of Secondary Education in the United States," 372–374.

56. Angus and Mirel, "Presidents, Professors, and Lay Boards of Education," 26–27; Tyack, *One Best System;* David Tyack and Elisabeth Hansot, *Managers of Virtue: Public School Leadership in American, 1820–1980* (New York: Basic Books, 1982).

57. Angus and Mirel, "Presidents, Professors, and Lay Boards of Education," 26–27.

58. "Discussion of the Report of the Committee on College Entrance Requirements in Joint Session of Secondary and Higher Departments," in *Journal of Proceedings and Addresses of the Thirty-Ninth Annual Meeting*, National Education Association (Published by the NEA, 1900), 451–452.

59. Ibid., 452.

60. National Education Association, *Report of the Committee on College Entrance Requirements, July, 1899*, 20.

61. Ibid., 42–44.

62. James M. Greenwood, "Discussion," *Journal of Education* 39 (March 8, 1894), 150.

63. Caskie Harrison, "Secondary School Studies," *Journal of Education* 30 (March 22, 1894), 180.

64. M. A. Whitney, "A Uniform Course of Study for All the Schools of the State," in *Fifty-Eighth Annual Report of the Superintendent of Public Instruction of the State of Michigan* (Lansing: Robert Smith & Co., 1895), 147–148.

65. Charles W. Eliot, "The Unity of Educational Reform," *Educational Review* 8 (October 1894), 214, 223.

66. National Education Association, *Report of the Committee of Ten on Secondary School Studies*, 53.

67. Ibid., 54–55.

68. James B. Angell, *The Reminiscences of James Burrill Angell* (New York: Longmans, Green, and Co., 1912), 245–246; University of Michigan, *The President's Report to the Board of Regents for the Year Ending September 30, 1888* (Ann Arbor: Published by the University, 1888), 7.

69. University of Michigan, *The President's Report to the Board of Regents for the Year Ending June 30, 1879* (Ann Arbor: Published by the University, 1879), 10–11.

70. Angell, *Reminiscences of James Burrill Angell*, 245–246.

71. "Letter to Paul H. Hanus, from Charles W. Eliot, March 25, 1891," Letter Press Volume, 1889–1898, Charles W. Eliot Papers, Harvard University Archives, Harvard University, 15.

72. University of Wisconsin, *Biennial Report of the Regents of the University for the Years 1894–95 and 1895–96* (Madison, WI: Democrat Printing Company, State Printer, 1896), 26.

73. University of Wisconsin, *Fourth Biennial Report of the Board of Regents of the University of Wisconsin for the School Years 1888–9, 1889–90* (Madison, WI: Democrat Printing Company, State Printers, 1890), 42.

74. Leslie A. Butler, *The Michigan Schoolmasters' Club: A Story of the First Seven Decades, 1886–1957* (Ann Arbor: University of Michigan, 1958), 82–84.

75. Harvard College, *Annual Reports of the President and Treasurer of Harvard College, 1872–73* (Cambridge: University Press, 1874), 20.

76. University of Wisconsin, *Biennial Report of the Regents of the University for the Years 1896–97 and 1897–98*, University of Wisconsin (Madison, WI: Democrat Printing Company, State Printer, 1898), 9–10.

77. W. D. MacClintock, "University College," in *The President's Report, July, 1892–July, 1902*, University of Chicago (Chicago: The University of Chicago Press, 1903), 146, 152–155; Edmund J. James, "The College for Teachers," in *The President's Report, July, 1898–July, 1899*, University of Chicago (Chicago: The University of Chicago Press, 1900), 79–80.

78. Nathaniel Butler, Frank J. Miller, and Charles R. Barnes, "The Department of University Relations," in *The President's Report, July, 1902–July, 1904*, University of Chicago (Chicago: The University of Chicago Press, 1905), 176.

79. Nicholas Murray Butler, "The Duty of the University to the Teaching Profession," *Journal of Education* 32 (August 21, 1890), 118–119.

80. "The May Conference at the University," *Wisconsin Journal of Education* 30 (June 1900), 127.

81. Ibid., 127; University of Wisconsin, *Biennial Report of the Regents of the University for the Years 1898–99 and 1899–1900* (Madison, WI: Democrat Printing Company, State Printer, 1900), 17; University of Wisconsin, *Biennial Report of the Regents of the University for the Years 1900–01 and 1901–02* (Madison, WI: Democrat Printing Company, State Printer, 1902), 14–15.

# 6    Regional Efforts and a Renewed Focus on National Reform

1. Carnegie Foundation for the Advancement of Teaching, *Fourth Annual Report of the President and Treasurer* (New York: Carnegie Foundation for the Advancement of Teaching, 1909), 134.

2. Edward A. Krug, *The Shaping of the American High School, 1880–1920* (Madison: The University of Wisconsin Press, 1969), 145; Nicholas Murray Butler, "Uniform College Entrance Requirements with a Common Board of Examiners," in *Proceedings of the Thirteenth Annual Convention,* Association of Colleges and Preparatory Schools of the Middle States and Maryland (Albany: University of the State of New York, 1900), 46–47.

3. Charles W. Eliot, "Discussion," in *Addresses and Proceedings of the Preliminary Meeting, Oct. 16 and 17, 1885, First Annual Meeting, Oct. 16, 1886, and First Special Meeting, Jan. 7 and 8, 1887,* New England Association of Colleges and Preparatory Schools (Boston, Mass: Published by the Association, 1892), 16.

4. Butler, "Uniform College Entrance Requirements with a Common Board of Examiners," 46–47; John A. Valentine, *The College Board and the School Curriculum: A History of the College Board's Influence on the Substance and Standards of American Education, 1900–1980* (New York: College Entrance Examination Board, 1987), 3–4, 8–10.

5. "Meeting of the Association of Colleges in New England at Amherst, Nov. 1st, 1894," Letter Press Volume, 1889–1898, Charles W. Eliot Papers, Harvard University Archives, Harvard University, 84a–85.

6. Nicholas Murray Butler, "Entrance to College," *New York Times* (December 17, 1899), 25.

7. Butler, "Uniform College Entrance Requirements with a Common Board of Examiners," 46; George Herbert Locke, "Editorial Notes," *School Review* 10 (November 1902), 711.

8. Julius Sachs, "Discussion," in *Proceedings of the Thirteenth Annual Convention,* Association of Colleges and Preparatory Schools of the Middle States and Maryland (Albany: University of the State of New York, 1900), 138–139.

9. Christopher Gregory, "Discussion," in *Proceedings of the Thirteenth Annual Convention,* Association of Colleges and Preparatory Schools of the Middle States and Maryland (Albany: University of the State of New York, 1900), 61.

10. Nicholas Murray Butler, *Across the Busy Years: Recollections and Reflections* (New York: Charles Scribner's Sons, 1939), 198.

11. Nicholas Murray Butler, "Establishment of a College Entrance Examination Board for the Middle States and Maryland," in *Proceedings of the Fourteenth Annual Convention,* Association of Colleges and Preparatory Schools of the Middle States and Maryland (Albany: University of the State of New York, 1901), 48–49; College Entrance Examination Board, *Second Annual Report of the Secretary, 1902* (New York: Published by the Board, 1902), 4–5. The first twelve colleges and universities to join were Barnard College, Bryn Mawr College, Columbia University, Cornell University, Johns Hopkins University, New York University, Rutgers College, Swarthmore College, Union College, University of Pennsylvania, Vassar College, and Woman's College of Baltimore.

12. Butler, "Establishment of a College Entrance Examination Board for the Middle States and Maryland," 51; College Entrance Examination Board, *Eighth Annual Report of the Secretary, 1908* (New York: Published by the

Board, 1908), 2; Claude M. Fuess, *The College Board: Its First Fifty Years* (New York: Columbia University Press, 1950), 60–61; Harold Wechsler, *The Qualified Student: A History of Selective College Admission in America* (New York: John Wiley & Sons, 1977), 102–103.

13. "Editorial: The Problem of College Admission," *Educational Review* 21 (January 1901), 106–107.

14. College Entrance Examination Board, *Second Annual Report of the Secretary, 1902*, 37.

15. Harold Wechsler, "Eastern Standard Time: High School–College Collaboration and Admission to College, 1880–1930," in *A Faithful Mirror: Reflections on the College Board and Education in America,* edited by Michael C. Johanek (College Entrance Examination Board, 2001), 45.

16. College Entrance Examination Board, *Second Annual Report of the Secretary, 1902.*

17. Ibid., 10.

18. Clement L. Smith, quoted in Nicholas Murray Butler, "College Entrance Examination Board of the Middle States and Maryland: First Annual Report of the Secretary on the Examinations of 1901" *Educational Review* 22 (October 1901), 4.

19. Butler, "College Entrance Examination Board of the Middle States and Maryland: First Annual Report of the Secretary on the Examinations of 1901," 3, 8–14; Fuess, *College Board,* 34, 48–52; Edward L. Harris, "The Public High School: Its Status and Present Development," in *Proceedings of the Twelfth Annual Meeting of the North Central Association of Colleges and Secondary Schools* (Urbana, IL: Published by the Association, 1907), 16.

20. College Entrance Examination Board, *Second Annual Report of the Secretary, 1902,* 2, 5. Although Yale refused to accept the Board's certificates without first reexamining the results, the Sheffield Scientific School at Yale accepted the results without any further review.

21. College Entrance Examination Board, *Fourth Annual Report of the Secretary, 1904* (New York: Published by the Board, 1904), 2; College Entrance Examination Board, *Fifth Annual Report of the Secretary, 1905* (New York: Published by the Board, 1905), 1; College Entrance Examination Board, *Ninth Annual Report of the Secretary, 1909* (New York: Published by the Board, 1909), 1; College Entrance Examination Board, *Tenth Annual Report of the Secretary, 1910* (New York: Published by the Board, 1910), 1, 10; Valentine, *College Board and the School Curriculum,* 18; Sidney Marland, Jr., *The College Board and the Twentieth Century* (New York: College Entrance Examination Board, 1975), 5.

22. College Entrance Examination Board, *Fifth Annual Report of the Secretary, 1905,* 8, 14; College Entrance Examination Board, *Fourth Annual Report of the Secretary, 1904,* 2; Valentine, *College Board and the School Curriculum,* 18; Marland, *College Board and the Twentieth Century,* 5; North Central Association of Colleges and Secondary Schools, *Proceedings of the Ninth Annual Meeting of the North Central Association of Colleges and Secondary Schools* (Columbus, Ohio: Published by the Association, 1904), 61–62.

23. Fuess, *College Board,* 61–62.
24. Carnegie Foundation for the Advancement of Teaching, *Second Annual Report of the President and Treasurer* (New York: Carnegie Foundation for the Advancement of Teaching, 1907), 74.
25. Krug, *Shaping of the American High School,* 150–151.
26. Wilson Farrand, "Five Years of The College Entrance Examination Board," *Educational Review* 30 (October 1905), 225.
27. Nicholas Murray Butler, "College Entrance Examination Board of the Middle States and Maryland: First Annual Report of the Secretary on the Examinations of 1901" *Educational Review,* 22 (October 1901), 2; "Document No. 2," reprint, College Entrance Examination Board, *The Work of the College Entrance Examination Board, 1901–1925: The Solution of Educational Problems Through the Cooperation of All Vitally Concerned* (Boston: Ginn and Company, 1926), 73.
28. College Entrance Examination Board, *Seventh Annual Report of the Secretary, 1907* (New York: Published by the Board, 1907), 1–2.
29. Ibid., 1.
30. College Entrance Examination Board, *Second Annual Report of the Secretary, 1902,* 10, 22; College Entrance Examination Board, *Ninth Annual Report of the Secretary, 1909,* 17, 43.
31. Marvin Lazerson, "The College Board and American Educational History," in *A Faithful Mirror,* 382; Andrea Walton, "Cultivating a Place for Selective All-Female Education in a Coeducational World: Women Educators and Professional voluntary Associations, 1880–1926," in *A Faithful Mirror,* 155, 161–164; College Entrance Examination Board, *Eighth Annual Report of the Secretary, 1908,* 15–17. The seven female colleges were Smith, Bryn Mawr, Wellesley, Vassar, Mt. Holyoke, Barnard, and Woman's College of Baltimore.
32. "Admission to College on Certificate of Secondary Schools," in *Report of the Commissioner of Education for the Year 1902,* United States Bureau of Education (Washington, D.C.: Government Printing Office, 1903), 527. The nine colleges were Amherst College, Boston University, Brown University, Dartmouth College, Mount Holyoke College, Smith College, Tufts College, Wellesley College, and Wesleyan University.
33. Commission of Colleges in New England on Admission Examination, *Fourteenth Annual Report, 1899–1900* (Providence: Snow & Farnham, Printers, 1900), 10–11; Commission of Colleges in New England on Admission Examination, *Fifteenth Annual Report, 1900–1901* (Providence: Snow & Farnham, Printers, 1901), 7–10.
34. Commission of Colleges in New England on Admission Examination, *Fifteenth Annual Report, 1900–1901,* 13–18.
35. "Admission to College on Certificate of Secondary Schools," 527–528.
36. Ibid., 528.
37. Nathaniel F. Davis, "Is the Present Mode of Granting Certificate Rights to Preparatory Schools Satisfactory?," in *Official Report of the Twenty-First Annual Meeting of the New England Association of Colleges and Preparatory Schools* (Chicago: Reprinted from *School Review,* 1906), 74.

38. College Entrance Examination Board, *Fifth Annual Report of the Secretary, 1905*, 8–9, 26–28; College Entrance Examination Board, *Eighth Annual Report of the Secretary, 1908*, 10, 15–17.

39. "The Unification of Requirements for Admission to College," *The University (of Michigan) Record* 4 (February 1895), 93; Mark Newman, *Agency of Change: One Hundred Years of the North Central Association of Colleges and Schools* (Kirksville, Missouri: Thomas Jefferson University Press, 1996), x–xvi, 22–25, 45–47; Calvin Olin Davis, *A History of the North Central Association of Colleges and Secondary School, 1895–1945* (Ann Arbor: The North Central Association of Colleges and Secondary Schools, 1945), 3–5, 8–15, 45.

40. Davis, *History of the North Central Association of Colleges and Secondary School*, 45–47; Allen S. Whitney, "The Problem of Harmonizing State Inspection by Numerous Colleges so as to Avoid Duplication of Work and Secure the Greatest Efficiency," in *Proceedings of the Sixth Annual Meeting of the North Central Association of Colleges and Secondary Schools* (Ann Arbor: Published by the Association, 1901), 25–26.

41. North Central Association of Colleges and Secondary Schools, *Proceedings of the Sixth Annual Meeting of the North Central Association of Colleges and Secondary Schools* (Ann Arbor: Published by the Association, 1901), 70–71.

42. Ibid., 70–71.

43. North Central Association of Colleges and Secondary Schools, *Proceedings of the Seventh Annual Meeting of the North Central Association of Colleges and Secondary Schools* (Ann Arbor: Published by the Association, 1902), 8; Davis, *History of the North Central Association of Colleges and Secondary School*, 37–38.

44. North Central Association of Colleges and Secondary Schools, *Proceedings of the Seventh Annual Meeting*, 7–8.

45. Ibid., 8.

46. Ibid., 35–36; North Central Association of Colleges and Secondary Schools, *Proceedings of the Ninth Annual Meeting*, 47; North Central Association of Colleges and Secondary Schools, *Proceedings of the Twelfth Annual Meeting of the North Central Association of Colleges and Secondary Schools* (Urbana, IL: Published by the Association, 1907), 57.

47. North Central Association of Colleges and Secondary Schools, *Proceedings of the Seventh Annual Meeting*, 35–36; North Central Association of Colleges and Secondary Schools, *Proceedings of the Ninth Annual Meeting*, 47; North Central Association of Colleges and Secondary Schools, *Proceedings of the Twelfth Annual Meeting*, 57.

48. North Central Association of Colleges and Secondary Schools, *Proceedings of the Seventh Annual Meeting*, 36–38, 42.

49. Davis, *History of the North Central Association of Colleges and Secondary School*, 54–55; "Admission to College on Certificate of Secondary Schools," 529–531; North Central Association of Colleges and Secondary Schools, *Proceedings of the Eighth Annual Meeting of the North Central Association of Colleges and Secondary Schools* (Ann Arbor: Published by the Association, 1903), 86.

50. North Central Association of Colleges and Secondary Schools, *Proceedings of the Seventh Annual Meeting*, 43.

51. North Central Association of Colleges and Secondary Schools, *Proceedings of the Eighth Annual Meeting*, 63, 65.

52. Ibid., 67–68.

53. Harry Pratt Judson, "The Outlook for the Commission," in *Proceedings of the Ninth Annual Meeting of the North Central Association of Colleges and Secondary Schools* (Columbus, Ohio: Published by the Association, 1904), 59–60.

54. George N. Carman, "Shall We Accredit Colleges?," in *Proceedings of the Eleventh Annual Meeting of the North Central Association of Colleges and Secondary Schools* (Columbus, Ohio: Published by the Association, 1906), 96.

55. E. L. Coffen, "The Inspection and Accrediting of Colleges and Universities," in *Proceedings of the Eleventh Annual Meeting of the North Central Association of Colleges and Secondary Schools* (Columbus, Ohio: Published by the Association, 1906), 97, 99.

56. North Central Association of Colleges and Secondary Schools, *Proceedings of the Fourteenth Annual Meeting of the North Central Association of Colleges and Secondary Schools* (Chicago: Published by the Association, 1909), 52–53, 58; North Central Association of Colleges and Secondary Schools, *Proceedings of the Fifteenth Annual Meeting of the North Central Association of Colleges and Secondary Schools* (Chicago: Published by the Association, 1910), 76.

57. Krug, *Shaping of the American High School*, 159–160.

58. North Central Association of Colleges and Secondary Schools, *Proceedings of the Eighteenth Annual Meeting of the North Central Association of Colleges and Secondary Schools* (Published by the Association, 1913), 63–65.

59. Association of Colleges and Preparatory Schools of the Southern States, *Proceedings of the First Meeting* (Association of Colleges and Preparatory Schools of the Southern States, 1895), 3–7; Krug, *Shaping of the American High School*, 127; Donald C. Agnew, *Seventy–Five Years of Educational Leadership* (Atlanta: Southern Association of Colleges and Schools, 1970), 2, 5; Edwin Mims, *Chancellor Kirkland of Vanderbilt* (Nashville: Vanderbilt University Press, 1940), 129–132; George Herbert Locke, "Editorial Notes: A Significant Forward Movement in Secondary Education in the South," *School Review* 13 (March 1905), 263; Association of Colleges and Preparatory Schools of the Southern States, *Proceedings of the Eleventh Annual Meeting* (Chattanooga, Tenn: Press of Southern Educational Review, n.d.), iv–v. The initial six colleges and universities were Vanderbilt University, the University of North Carolina, the University of Mississippi, Washington and Lee University, the University of the South, and Trinity College (Duke).

60. Jabez L. M. Curry, "Education in the Southern States," in *Proceedings of the Second Capon Springs Conference for Christian Education in the South* (Capon Springs Conference for Christian Education in the South, 1899), 28.

61. Association of Colleges and Preparatory Schools of the Southern States, *Proceedings of the Ninth Annual Meeting* (Printed by Brandon Printing Company, Nashville, Tenn., n.d.), 20; Paul H. Saunders, "Report of the Committee on Uniform Entrance Examinations," in *Proceedings of the Tenth*

*Annual Meeting*, Association of Colleges and Preparatory Schools of the Southern States (Chattanooga, Tenn: Press of Southern Educational Review, n.d), 44, 47; F. W. Moore, "Report of the Williamstown Conference on Admission to College," in *Proceedings of the Twelfth Annual Meeting*, Association of Colleges and Preparatory Schools of the Southern States (Nashville, Tenn: Press of Marshall & Bruce Company, n.d.), 11.

62. Saunders, "Report of the Committee on Uniform Entrance Examinations," 47–48.

63. Association of Colleges and Preparatory Schools of the Southern States, *Proceedings of the Eleventh Annual Meeting*, 2; Paul H. Saunders, "Our Experiment in Uniform Examinations," in *Proceedings of the Eleventh Annual Meeting*, Association of Colleges and Preparatory Schools of the Southern States (Chattanooga, Tenn: Press of Southern Educational Review, n.d.), 32–33, 35. North Carolina and Virginia statements quoted in Saunders, "Our Experiment in Uniform Examinations," 32–33.

64. Association of Colleges and Preparatory Schools of the Southern States, *Proceedings of the Sixteenth Annual Meeting* (Nashville, Tenn: Press of Standard Printing Company, 1910), 26; Association of Colleges and Preparatory Schools of the Southern States, *Proceedings of the Seventeenth Annual Meeting* (Nashville, Tenn: Publishing House of the Methodist Episcopal Church, n.d.), 21.

65. College statements quoted in Paul H. Saunders, "The Report of the Uniform Entrance Examination Committee," in *Proceedings of the Twelfth Annual Meeting*, Association of Colleges and Preparatory Schools of the Southern States (Nashville, Tenn: Press of Marshall & Bruce Company, n.d.), 56.

66. Saunders, "Report of the Uniform Entrance Examination Committee," 56–57.

67. Association of Colleges and Preparatory Schools of the Southern States, *Proceedings of the Eighteenth Annual Meeting* (Nashville, Tenn: Publishing House of the Methodist Episcopal Church, n.d.), 23, 32; Agnew, *Seventy-Five Years of Educational Leadership*, 9.

68. Agnew, *Seventy-Five Years of Educational Leadership*, 8–9; United States Bureau of Education, *Report of the Commissioner of Education for the Year 1899–1900* (Washington, D.C.: Government Printing Office, 1901), 2129, 2130, 2145, 2162; United States Bureau of Education, *Report of the Commissioner of Education for the Year Ended June 30, 1911* (Washington, D.C.: Government Printing Office, 1912), 1192, 1196, 1205.

69. Association of Colleges and Preparatory Schools of the Southern States, *Proceedings of the Seventeenth Annual Meeting*, 23–24; Agnew, *Seventy-Five Years of Educational Leadership*, 8–9; Association of Colleges and Preparatory Schools of the Southern States, *Proceedings of the Eighteenth Annual Meeting*, 30.

70. Association of Colleges and Preparatory Schools of the Southern States, *Proceedings of the Seventeenth Annual Meeting*, 23–24; Agnew, *Seventy-Five Years of Educational Leadership*, 8–10, 28–29; Leland Stanford Cozart, *A History of the Association of Colleges and Secondary Schools, 1934–1965* (Charlotte, NC: Heritage Printers, Inc., 1967), 1–2.

71. Association of Colleges and Preparatory Schools of the Southern States, *Proceedings of the Eighteenth Annual Meeting*, 29.

72. United States Bureau of Education, *Report of the Commissioner of Education for the Year Ended June 30, 1911*, 1184, 1224.

73. National Association of State Universities, *Transactions and Proceedings* (Published by the Association, 1905), 9, 75–76.

74. George Edwin MacLean, "An American Federation of Learning," in *Proceedings of the Eleventh Annual Meeting of the North Central Association of Colleges and Secondary Schools* (Columbus, Ohio: Published by the Association, 1906), 9.

75. National Association of State Universities, *Transactions and Proceedings* (Bangor, Maine: Bangor Co-Operative Printing, Co, 1907), 14–15; North Central Association of Colleges and Secondary Schools, *Proceedings of the Thirteenth Annual Meeting of the North Central Association of Colleges and Secondary Schools* (Chicago: Published by the Association, 1908), 44.

76. National Conference Committee on Standards of Colleges and Secondary Schools, *Minutes of the Conference* (Published by the Association, 1909), 3; National Association of State Universities, *Transactions and Proceedings* (Hamilton, Ohio: Republican Publishing Company, 1910), 257–261; College Entrance Examination Board, *Tenth Annual Report of the Secretary, 1910*, 3; North Central Association of Colleges and Secondary Schools, *Proceedings of the Fifteenth Annual Meeting*, 31–34.

77. Krug, *Shaping of the American High School*, 160–161; Ellen Condliffe Lagemann, *Private Power for the Public Good: A History of the Carnegie Foundation for the Advancement of Teaching* (Middletown, CT: Wesleyan University Press, 1983), 37–39; Henry S. Pritchett, "Scope and Practical Workings of the Carnegie Foundation for the Advancement of Teaching," *Journal of Education* 68 (December 17, 1908), 657; Carnegie Foundation for the Advancement of Teaching, *First Annual Report of the President and Treasurer* (New York: Carnegie Foundation for the Advancement of Teaching, 1906), 15–16.

78. Carnegie Foundation for the Advancement of Teaching, *First Annual Report of the President and Treasurer*, 20, 79.

79. Carnegie Foundation for the Advancement of Teaching, *Second Annual Report of the President and Treasurer*, 69; North Central Association of Colleges and Secondary Schools, *Proceedings of the Twelfth Annual Meeting*, 62–63; Lagemann, *Private Power for the Public Good*, 95; College Entrance Examination Board, *Seventh Annual Report of the Secretary, 1907*, 7.

80. Krug, *Shaping of the American High School*, 160–161; Lagemann, *Private Power for the Public Good*, 95; Howard J. Savage, *Fruit of an Impulse: Forty-Five Years of the Carnegie Foundation, 1905–1950* (New York: Harcourt, Brace and Company, 1953), 66, 102; Carnegie Foundation for the Advancement of Teaching, *First Annual Report of the President and Treasurer*, 38–39, 47; Carnegie Foundation for the Advancement of Teaching, *Second Annual Report of the President and Treasurer*, 67.

81. Carnegie Foundation for the Advancement of Teaching, *First Annual Report of the President and Treasurer,* 24–25; Carnegie Foundation for the Advancement of Teaching, *Fifth Annual Report of the President and of the Treasurer* (New York: Carnegie Foundation for the Advancement of Teaching, 1910), 31–32.

82. J. H. Kirkland, "Requirements for Admission to College," in *Proceedings of the Thirteenth Annual Meeting,* Association of Colleges and Preparatory Schools of the Southern States (Nashville, Tenn: Press of Marshall & Bruce Company, n.d.), 69

83. Carnegie Foundation for the Advancement of Teaching, *First Annual Report of the President and Treasurer,* 24–25; Carnegie Foundation for the Advancement of Teaching, *Fifth Annual Report of the President and of the Treasurer,* 31–32.

84. Carnegie Foundation for the Advancement of Teaching, *Third Annual Report of the President and Treasurer* (New York: Carnegie Foundation for the Advancement of Teaching, 1908), 62–63; "The Carnegie Pension," *Springfield Republican,* reprint, *Journal of Education* 71 (April 28, 1910), 459; Charles C. Heyl, "The Carnegie Foundation and Some American Educational Problems," *Journal of Education* 71 (May 26, 1910), 564–565.

85. Carnegie Foundation for the Advancement of Teaching, *Second Annual Report of the President and Treasurer,* 66; National Association of State Universities, *Transactions and Proceedings* (Bangor, Maine: Bangor Co-Operative Printing, Co, 1909), 62–63; Krug, *Shaping of the American High School,* 161–162.

86. Carnegie Foundation for the Advancement of Teaching, *Second Annual Report of the President and Treasurer,* 63.

# Epilogue: Looking Ahead by Looking to the Past

1. James R. Angell, "The Endowed Institution of Higher Education—Its Relation to Public Education," in *Proceedings of the Sixty-Sixth Annual Meeting,* National Education Association (Washington, D.C.: The National Education Association, 1928), 774.

2. Bill Gates, "National Education Summit on High Schools," Bill and Melinda Gates Foundation, February 26, 2005: www.gatesfoundation.org/MediaCenter/Speeches/Co-ChairSpeeches/BillgSpeeches/BGSpeechNGA-050226.htm.

3. See William J. Reese, *The Origins of the American High School* (New Haven: Yale University Press, 1995).

# Index

Printed in the United States
By Bookmasters